Microstation v8

Simplified

First Edition

Corey C. Hutchinson

SOLUTIONS Technical Training Center, Inc.
Texas
© 2008

ISBN 978-0-9817788-0-8 (Volume 1)
Solutions Technical Training Center, Inc.
Houston, Texas 77042

Library of Congress Control Number: 2008904515

The use of the names Microstation, AutoCAD, Microsoft, Windows, and others are for reference only. Microstation, AutoCAD, Microsoft, Windows, and other references used in this book are the registered and licensed names of their respective owners, whom have no affiliation with the author or publisher of this book.

In memory of my grandparents,

Mr. Eustace and Mrs. Linda Willis

and ...

Preface

Microstation is a multi-functional and versatile Computer Aided Design (CAD) program, widely used by various technical industries to create accurate two-dimensional (2D) and three-dimensional (3D) designs/drawings. It can be considered the all-in-one program of the CAD industry. Some programs do 2D well, while others do well in 3D. Microstation on the other hand does both exceptionally well, providing the tools and features to create whatever the mind can conceptualize.

In _Microstation v8: Simplified_, Volume 1, the focus will be on the 2D aspects of the program. This is Part A of a multi-part volume. The objective of this text is to provide a good understanding of the tools to efficiently manipulate the program. Therefore, the book uses simple and straightforward methods to assist the reader in grasping the features of the program. It provides easy steps and procedures, along with explanations of the features and options.

The book begins with an introduction to the basic operation and setting up of the drawing. This starts with the reader learning how to create and open new files or open existing files. Explanation of the design plane (drawing area) is provided in order to understand the user interface, such as toolbars, menus, and status bar. Coordinate Systems are covered, allowing improved orientation in the drawing area. Unit Setup is demonstrated because it determines a drawing's form and size, a critical aspect for manufacturing. One of the most important features of Microstation is View Control, which controls the visibility of the drawing on the screen.

After the drawing is initially setup, the drawing, manipulation, and modification tools are discussed. The objects drawn in Microstation are also called elements. The drawing tools can be divided into two categories, linear and curved. Lines, rectangles, and polygons are examples of linear tools, while curved tools are circles, arcs, and ellipses. If the desired design results are not reached, manipulation and modification tools are used to correct it. Manipulation tools are ones that relocate or change the size of an element, not its form. Some of the manipulating tools are move, copy, rotate, and scale. Modification tools, on the other hand, changes the geometry of an object. Examples of the modifying tools are trim, extend, fillet, and chamfer.

Attributes (Level, Color, Style, and Weight) are the characteristics of an element, and the associated tools for setting and changing them are explained. The manipulation of these attributes is simple, and the understanding how to use these characteristics creates a clearer, easier to read drawing. Levels are one of the most important aspects of an organized drawing. Synonymous to transparent sheets of paper (that's drawn on and layered on top of each other), you can turn them on and off to less complicate the overall drawing. A good practice in CAD is to setup the Attributes first, in preparation for their eventual use in the drawing, before any element/object is drawn.

Texts are individual characters, words, and sentences. The text creating tools are explained, along with text editing tools. Regular single and multi line text are available, along with Notes. Tools are also available to edit, match, and change text.

Printing is the last feature explained in Part A of _Microstation v8: Simplified_, Volume 1. This section discusses printing parameters, such as the print scale. Drawing in Microstation is usually done full scale, but any scaling concern beyond that point is dealt with the print scale. Here the drawing is adjusted to the paper scale, allowing for proper interpretation of what was drawn.

Nonetheless, let's begin.

Table of Contents

Chapter 1: Introduction

Without a beginning, there is no end. This chapter begins with the basics of getting started in Microstation v8. The objective is to provide introductory information on the working interfaces and initial setup of the program. These are the essentials of every drawing. It provides the foundation for the drawing and makes it an easier process to derive the desired output. After completing this chapter, you'll be able to do the following:

- Open, save, and close a new or existing drawing
- Get help on topics that are unfamiliar
- Know the different parts of the Design Plane
- Know the Coordinate System used for 2D drawings
- Setup units for your drawing
- Use Grids, Snaps, and Key-ins to complete a precise drawing
- Use Edits to go back and forth in your drawing

I.   Starting Microstation
a.   Create and Open a New Drawing
When Microstation starts, it provides a prompt via the Microstation Manager, where preliminary settings are available to begin a design/drawing:

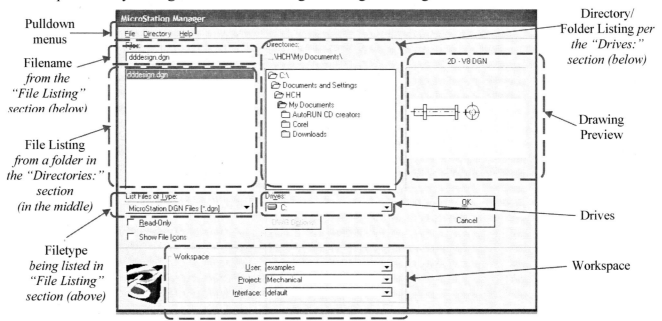

This area allows for the management and control of settings associated with drives, directories, and files. The Microstation Manager is analogous to Windows Explorer.

The **File** pulldown menu control features related to files as follows:
- New – creates a new file from scratch
- Copy – copy an existing file to a new one
- Rename – renames an existing file to a new name
- Delete – deletes an existing file
- Properties – inputs and/or displays information related to a file, such as author, company, and number of objects in a file.
- Merge – merges an existing file into another
- Compress – compresses an existing file to a smaller compact filesize
- Upgrade Files to v8 – upgrades an older version Microstation file to version 8.
  NOTE: *the File pulldown menu will also list recently viewed files*

The **Directory** pulldown menu control features related to directories as follows:
- New – creates a new directory
- Copy – copy an existing directory to a new one
- Compress – compresses files in an existing directory to smaller compact filesizes
- Upgrade Files to v8 – upgrades older version Microstation files in a directory to version 8.

The **Help** pulldown menu provides helpful information on the program and contacts for product support.

Before starting a drawing, one important feature of Microstation to understand is "Workspace". Workspace determines the characteristics/setup of the working environment for the drawing. It is representative of the type of drawing to be created (i.e.: General, Architectural, Mechanical, and etc.). Workspace will be covered more in-depth in *Microstation v8: Simplified*, Vol. 1, Part B. Nonetheless, an introduction is necessary.

Settings for Workspace are located on the bottom section of the Microstation Manager (see previous picture), and is made up of three components/parts:

1. User
   - Configures the drawing for a specific user, like a company division or contract.
   - If you change this option, the other two components listed below will change accordingly

2. Project
   - Configures the associated folders/files per the selection
   - Uses or creates a main folder per the selection, then assigns subfolders such as:
     - dgn: *to hold drawings created for the project*
     - out: *to hold converted/exported file formats for the project*
     - seed: *to hold seed files (templates) for the project*
     NOTE: *Example Projects are Architecture, Civil, Mapping, Mechanical, and etc.*

3. Interface
   - Changes feel and look of the Microstation tools (Main Palette and etc.)

The Microstation program is typically equipped with example Workspaces to choose from, which is accessed by setting the **User** component to "examples". Consequently, the **Project** component will list example projects to choose from. The **Interface** component should be set to "default".

For practical purposes at this introductory level, the Workspace section of the Microstation Manager should be set to the following selections before proceeding:

- User: *examples*

- Project: *General*

- Interface: *default*

If those options are not available, make the following alternative selections before proceeding:
- User: *untitled*

- Project: *untitled*

- Interface: *default*

NOTE: *the previously mentioned Workspace settings are recommended, even though they may differ from what may be depicted hereafter in this book.*

To start a new design, go to the **File** pulldown menu> **New** (at the top left corner of the Microstation Manager window), which will open the following window:

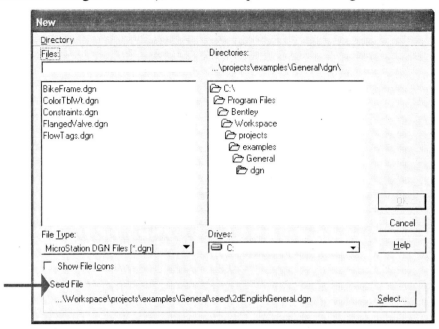

A seed file will be required for a new drawing. A seed file is a template drawing that has initial default drawing settings, such as the units, dimension, and text style, dependent on the type of drawing to be created (for example: a generic 2D design can use a seed2d.dgn, while a generic 3D design may use a seed3d.dgn. Additionally, a 2DEnglishGeneral.dgn or 2DEnglishMechanical.dgn seed file has settings for a "general" or "mechanical" 2D drawing with English units, respectively). The current seed file available for the drawing is shown at the bottom of the dialog box (see arrow on picture above). Alternate seed files can be chosen by going to **Select ...** at the bottom right corner of the "**New**" window, which will open the following window:

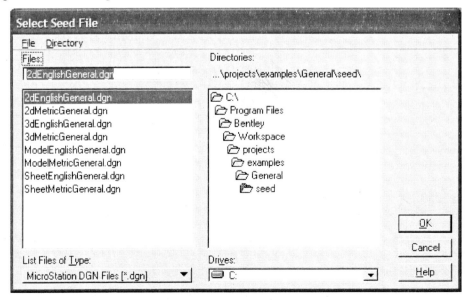

Choose a seed file and press **OK** to return back to the "**New**" window.

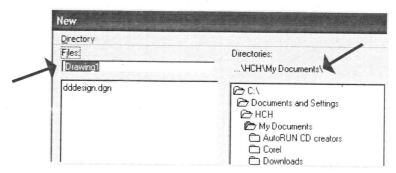

Choose or verify the appropriate directory/folder (see the arrow to the right on the picture above) in which the new design/drawing will be saved in. Enter the name of new file (see the arrow to the left on picture above).

Press **OK** to return back to the Microstation Manager. The newly created file will be shown in the file listing area of the Microstation Manager (within the directory/folder mentioned above). Choose the new file from the listing and press **OK** to open the file, which will display the drawing area as follows:

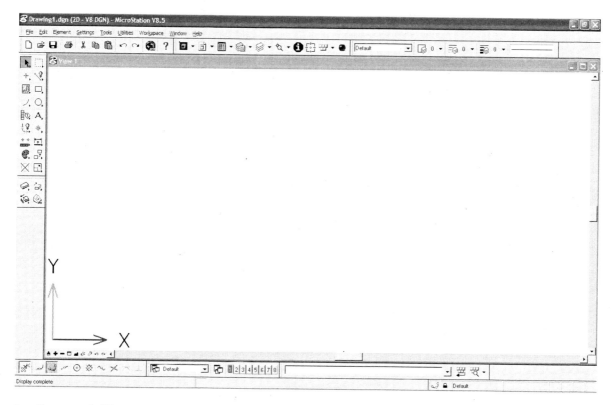

b.   Save and Close
Before saving a file, any changes to the current settings (for example: units and text size) should be saved (for the next time the drawing is re-opened), and the design/drawing compressed (to clear all excess areas and improve hard disk space used by the design/drawing) by going to the **File** pulldown menu> **Save Settings** and/or **File** pulldown menu> **Compress Design**, respectively. To avoid doing this every time, Microstation can

be set to do this automatically by going to the **Workspace** pulldown menu> **Preferences** > **Operation**, and placing a check by the associated options:

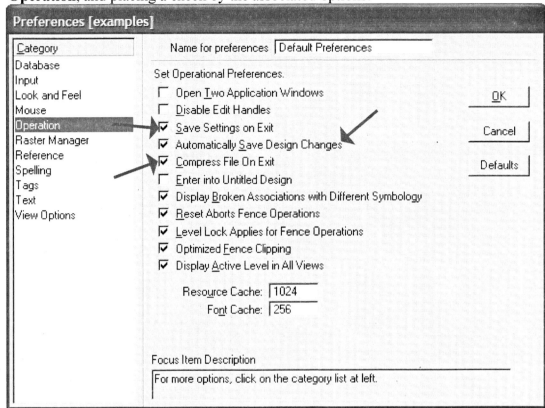

NOTE: *Check "Immediately Save Design Changes" to prevent any inadvertent loss of work, if a program failure occurs.*

Save the design by going to the **File** pulldown menu> **Save** (or 💾 icon, or press **Ctrl** and **S** on the keyboard)

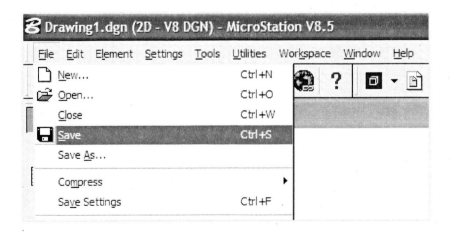

Use the **File** pulldown menu> **Save As** to save the file in a different name, different location than current directory, or different file type or version (see arrow in the next picture):

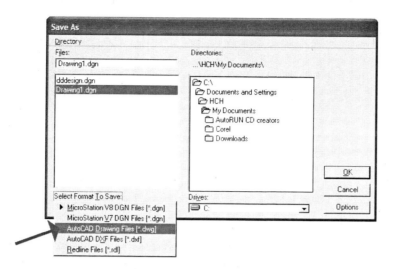

NOTE: *As shown in the filetype list, the drawing can be saved as an AutoCAD file. A brand new AutoCAD drawing can even be created in Microstation, and be worked on as an AutoCAD file from start through completion.*

Choose the **File** pulldown menu> **Close** in order to close the current design and go back to the Microstation Manager to have the opportunity to manipulate files and directories. This does not close the program. To do that, go to the **File** pulldown menu> **Exit**

c. Screen Display
    The microstation drawing area is made up of different functional parts, displayed on the screen, as shown in the following picture:

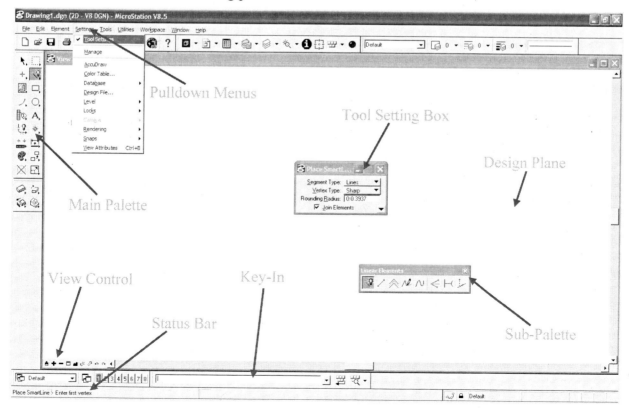

1. Design Plane – the area in which the design/drawing is completed

2. Pulldown Menu – Menus that allows for the access to tools and options in Microstation

3. Tool Setting Boxes – small windows that allows for the change of settings to the current tool being used in Microstation

4. Main Palette – an icon grouping of major tools used in Microstation

5. Sub-Palette – an icon grouping of sub-tools of major tools used in Microstation

6. Key-In – an area to input characters, words, or phrases that invokes commands for Microstation

7. Status Bar – an area displaying information about tools currently being used.

8. View Control – tools that control how the viewing area (views) is manipulated, such as zoom in, zoom out, and pan.

- Views (covered later in Chapter 2) are areas (windows) that display the design/drawing:

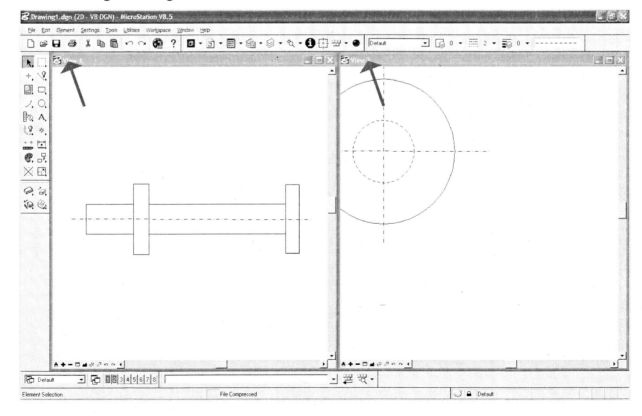

d. Getting Help

In order to get help on a command or feature of Microstation, go to the **Help** pulldown menu> **Contents**

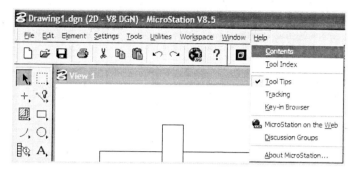

II. Command Windows

a. Design Plane

This is the drawing area. Everything drawn in the design plane will be drawn full scale 1 to 1 (1:1), exact size.

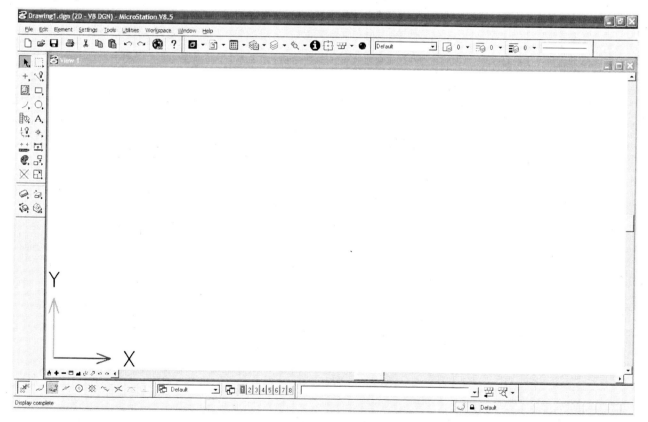

b. Coordinates System

Nominally, we consider everything we encounter in everyday life to be a three dimensional (3D) form, with every point on the object having three axes, X, Y, and Z. We identify an objects position and orientation by those three axes. Drawing on paper is drawing on a plane that only have two axes, X and Y (2D). In CAD, drawings are done in both 2D and 3D per the aforementioned axes, which form the Cartesian coordinate system. In this book, the focus is on 2D, ergo, the XY axes.

c. Global Origin and Coordinates
The center of the 2D paper/design plane is considered the Global Origin, which is specified as 0, 0 (X=0, Y=0). Going to the right from 0, 0 is going positive X (+X), while left is negative X (-X). Going up from 0, 0 is going positive Y (+Y), while down is negative Y (-Y).

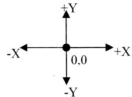

The Global origin can be re-specified be typing the Key-In GO=0, 0, then pick where the origin should be.

d. Main Palette
This is a grouping of commonly used draw, manipulation, and modification tools. They are listed below:

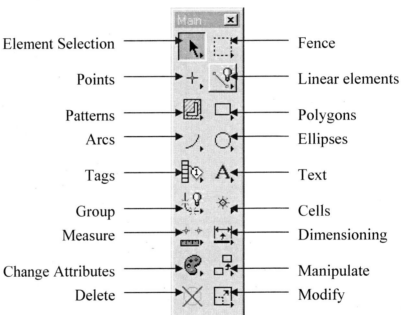

NOTE: *If the main palette is not visible, go to the **Tools** pulldown menu> **Main**> **Main***

i. Sub-Palette
1. These are sub-tools of the tools found in the main palette. There are too many sub-palettes to list here, however, if an icon (in the main palette) is picked with the "pick" mouse button (usually left button), and kept depressed while the mouse is dragged, the sub-palette will open. An example of the "Polygons" sub-palette is shown below:

NOTE: *A small arrow shown in the lower right corner of an icon indicates that a sub-palette is available, like*

e. Units Setup

Units control the size of the objects drawn, and therefore it is critical to the manufacture of the part. A part drawn with incorrect units can be costly. When setting up units in Microstation, two parameters needs to be adjusted: Working Units and Coordinate Readout.

  i. Working Units (defines the units to work with)

    1. These are the simple units to work with, such as feet and inches or meters and millimeters.

    2. They are usually set automatically when the seed file is chosen. They can be changed by going to the **Settings** pulldown menu> **Design File**> **Working Units**. The following dialog box will appear:

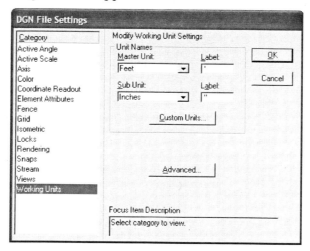

    3. They are made up of Master Units (i.e.: feet or meters) and Sub Units (i.e.: inches or millimeters), In other words, Sub-Units are the fractional part or sub-divided pieces of the Master Unit. Master Unit is also depicted as MU, while Sub-Unit is SU.

  ii. Coordinate Readout (defines how the user/designer and the program communicate "numbers" with each other)

    1. They are usually set automatically when the seed file is chosen. They can be changed by going to the **Settings** pulldown menu>**Design File**> **Coordinate Readout**. The following dialog box will appear:

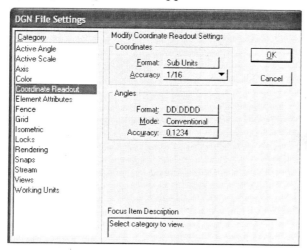

2. In the Coordinates area, the **Format** and **Accuracy** of the units inputted or displayed are chosen. The formats are
    - Working Units................ (MU:SU:PU)
    - Sub Units..................... (MU:SU)
    - Master Unit................... (MU)

    NOTE: *PU is precision unit, and is a division of Sub Unit. Working Units format is not recommended, as we never work in three different units, like yd:ft:in.*

    Format sets the lowest unit that the program will include in the display of the number. Accuracy sets the degree of correctness of the units. If fractional accuracy is needed, the choice is between 1/64 and 1/2. If decimal accuracy is needed, the choice is between no decimal place and 6 decimal places. For example, the program will display a measurement of a line that's 3-1/8 feet as follows (based upon the Working Units and Coordinate Readout setting shown):

Working Units set to:		*Coordinate Readout set to:*		*Program Displays Units as:*
Master Unit	**Sub Unit**	*Format*	*Accuracy*	
Feet	Inches	Master Unit	0.123 (Decimal)	3.125' or 3.125 (MU)
Feet	Inches	Master Unit	1/64 (Fractions)	3 1/8' or 3 1/8 (MU)
Feet	Inches	Sub Unit	0.123 (Decimal)	3' 1.500" or 3:1.500 (MU:SU)
Feet	Inches	Sub Unit	1/64 (Fractions)	3' 1 1/2" or 3:1 1/2 (MU:SU)
Feet	Inches	Working Units	0.1 (Decimal)	3' 1:12700.000" or 3:1:12700.000 (MU:SU:PU)
Feet	Inches	Working Units	1/64 (Fractions)	3' 1:12700" or 3:1:12700 (MU:SU:PU)

NOTE: *Always pay attention to the number of colons used in the display of the number to determine the current setting for the Coordinate Readout Format.*

3. In the Angles area, the **Format, Mode** and **Accuracy** of the angular units inputted or displayed are chosen. The angular formats are :
    - Decimal degrees (DD.DDDD)
    - Degrees, Minutes, and Seconds (DD MM SS)
    - Gradians
    - Radians.

    The angular modes are Conventional (typically used), Azimuth, and Bearing. The accuracy choice for angles is between no decimal place and 8 decimal places.

4. To make it simple, the **Working Units** panel controls which units to work with, but the **Coordinate Readout** panel controls how the unit is displayed. Dependent on the Coordinate Readout Format chosen, the program will look at the corresponding Working Unit dialog setting (Master Unit and/or Sub Unit) to see what it is set to, and display the number accordingly. A good rule of thumb is if the design being worked with uses one unit of measure like feet, set Master Unit to feet in the Working Units panel, and set the Coordinate Readout Format to Master Unit. Otherwise, specify the Master Unit and Sub Unit in the Working Units panel, and set the Coordinate Readout Format to Sub Unit. Remember, the Working Units Coordinate Readout Format is generally not used in most industries.

III. Precision Inputs
a. Grid

A Grid is like a matrix that acts as an aid in drawing accurately. They consists of equally spaced imaginary constraints/points (controlled by the user), in the X and Y axes. It is purely visual and cannot be printed.

   i. The grid can be turned ON and OFF, by going to the **Settings** pulldown menu> **View Attributes**> **Grid**.

The grid is shown as small crosses (plus signs) and dots.

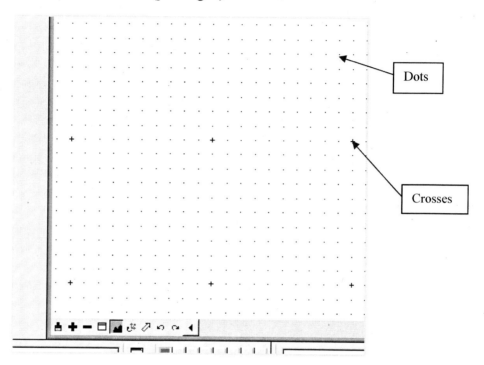

> **_NOTE:_** *If the grid can't be seen, the drawing may be zoomed out too far and/or the grid spacing is too small.*

   ii. Grid Setting

The spacing between the crosses and dots can be changed by going to the **Settings** pulldown menu> **Design File**> **Grid**.

Setting Grid Master defines the distance between the dots, and that between the cross and the closest dot. Setting the Grid Reference tells Microstation how many times to divide the dots between the crosses. The Grid Master times Grid Reference equals the size between the crosses. For example:

Working Units set to:		Coordinate Readout FORMAT set to::	Grid Master	Grid Reference	Distance between dots will be:	Distance between crosses will be:
Master Unit	Sub Unit					
Feet	Inches	Master Unit	0.25	4	3 inches	1 foot
Feet	Inches	Sub Unit	0:3	4	3 inches	1 foot

> **_NOTE:_** *The format in which the number is displayed in the Grid Master area is based upon that of the Coordinate Readout.*

iii. Grid Lock
Grid Lock allows for drawing only to the position of the crosses and dots. It can be accessed in five places:
   a. The lock icon on the status bar (see Tentative Snaps)
   b. **Settings** pulldown menu> **Design File> Grid> Grid Lock**
   c. **Settings** pulldown menu> **Locks**>
   d. **Settings** pulldown menu> **Locks> Full**
   e. **Settings** pulldown menu> **Locks> Toggles**

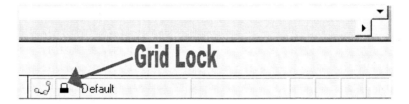

b. Snaps
Snaps allow for the picking of a specific location on an object/element to accurately complete the drawing. The Snap tools can be accessed in three places:
   i. The snap icon on the status bar

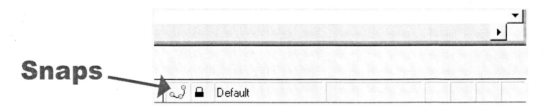

   ii. **Settings** pulldown menu> **Snaps> Button Bar**

The snap modes are:
- Nearest – nearest point on an object to the cursor
- Keypoint – key location points like endpoint, midpoint, and center point
- Midpoint – midpoint location on an object (each segment)
- Center – center point location on a object (any closed shape)
- Origin – local origin of an object (best used on text and cells)
- Bisector – the midpoint of the whole object (not segment)
- Intersection – intersection point of two objects
- Tangent – tangent point between an object and a circle, arc, or ellipse.
- Perpendicular - a point perpendicular between one object to the next object

NOTE: *These are the typical snap tools, but more are available. The other tool/modes can be accessed by simply positioning the cursor over the Snap toolbar, picking the right mouse button, and choose from the list.*

Snaps are sometimes called Tentative Snaps because they can be temporarily used when needed, however, there's always a default snap active. The default snap is depicted with a darkly shaded icon depressed in the Snap toolbar. Another snap mode can always be temporarily (tentatively) activated by picking it from Snap toolbar, which will be depicted in a lightly shaded icon depressed in the Snap toolbar.

   iii. **Settings** pulldown menu> **Locks**> **Full**

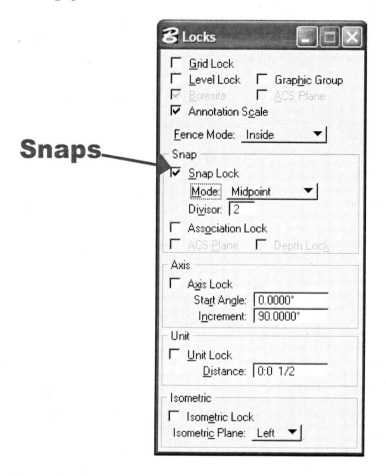

   iv. Button Assignment

In using the snap tools, the program sometimes provide assistance in snapping to objects. This occurs because of a feature called AccuSnap. This feature can be turned On and Off by pressing the first button in the Snap toolbar. The program will provide assistance with AccuSnap by highlighting the object of interest, and displaying a graphic of the snap mode at a certain location on the object. However, do not rely solely on the program to provide assistance in identifying the snap locations, because it may not pick what's needed at times. Therefore, use the mouse to identify the snap locations. This is accomplished by using one of the buttons on the mouse to pick a snap point on the object(s). The button assigned for this function may vary. To identify and/or change the button, go to the **Workspace** pulldown menu> **Button Assignments...**

The following dialog box will appear:

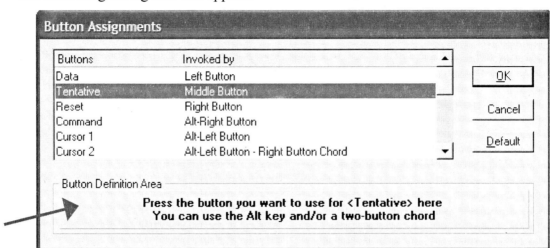

The Button Assignments dialog box lists available buttons, and what function is applied to or invoked by each. The Snap button is noted as "Tentative", and it is set to a button (Middle button, in this example). In order to change the button used for snapping, make sure Tentative is picked from the list of options, and then place the cursor in the "Button Definition Area". With the cursor in this area, press the desired button to use for snapping. The information noted for the Tentative button will change in the "Invoked by" column. Press OK to accept and close that dialog box.

Snaps are used with drawing and manipulation tools. It is assumed that when drawing a design, there is an expected result. Therefore, the designer has an idea of how objects will be laid out against one another. Snaps allows for accurately place objects against another. To snap to a specific location of an object with respect to another object, anytime during the drawing process, simply choose the appropriate snap mode. For example, to draw a line from the center of a circle to the midpoint of another line, the steps are as follows:

- Pick the draw tool (Line)
- Pick the snap mode (center snap)
- Press the snap button on the object (middle button on the circle)
- Press the accept button (left button) to accept the snapped location, or the reset button (right button) to cancel out of the snap mode
- If the accept button is picked, pick the next snap mode (midpoint snap)
- Press the snap button on the next object (middle button on the other line)
- Press the accept button (left button) to accept the snapped location, or the reset button (right button) to cancel out of the snap mode
- After the line has been drawn, cancel the command with the reset button

NOTE: *The "Accept"(Data) and "Reset" button may vary, dependent on the Button Assignments setting.*

To make it simple, every time the design requires snapping to an object, simply pick a snap mode, then accept or reject the snapped location. However, the process must always begin with a draw or manipulate operation.

c. Key-in

Key-ins are typed instructions used to invoke standard Microstation operations. They can be full words, or 2 letter commands called short-keys (i.e.: HATCH means Hatch Element Area or CO=3 means set active color to 3)

   i. Key-ins can be accessed from the **Utilities** pull-down menu> **Key-in**. Either the full Key-in dialog box, or the simple one located on the status bar, will be displayed. They are one in the same, just different dialog box configuration. Typing in the full Key-in dialog box will highlight the correct key-in in the keywords. *A second level keyword will be accessed, if available, to provide guidance through the proper steps to get to the end result.* For the simple key-in bar just enter the commands.

   ii. Microstation is not case sensitive, so key-ins can be upper and/or lowercase.

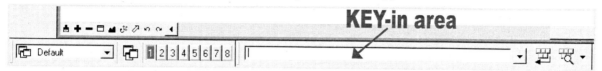

d. Element Selection Tool

The element selection tool is found in the main palette. The element selection is used to make an element active. It becomes highlighted, meaning it will be manipulated based upon the next tool that is picked. A crosshair or cube (sometimes called grips) will be placed at the vertices of the selected element. *If more than one element is selected, it is considered grouped, and they all will be manipulated.*

IV. Edits

a. Undo and Redo

Undo cancels the affect of the previous command, and Redo reverses the effect of the previous Undo command.

   i. Undo and Redo are available from the **Edit** pulldown menu or from the Standard toolbar (available from the **Tools** pulldown menu).

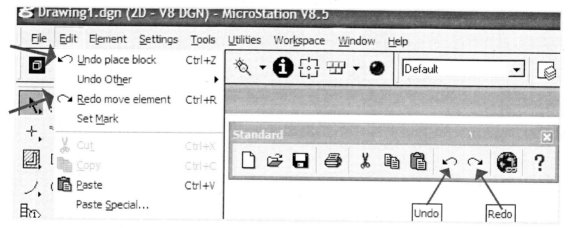

Chapter 2: Views

The eyes are the windows to the soul. In Microstation, the views are the windows to the drawing. Views allow for focusing on different areas and locations in the drawing. It allows you to zoom in and out of the drawing and move around in the drawing. Imagine a view as a display of what is seen through the lens of a camera, which is looking at your drawing. After completing this chapter, you'll be able to do the following:

- Open and close the view(s)
- Arrange and size the view(s)
- Control how the view(s) look
- Control attributes of the view(s)

I. View Settings

Views and windows are analogous terms in Microstation. A view is what is seen within the boundaries of the window. Microstation allows for a maximum of 8 windows, providing 8 views. These view windows can be manipulated and arranged. View windows can be opened and closed by going to the **Window** pulldown menu> **Views**> **Dialog** or placing a check beside specific views as shown below.

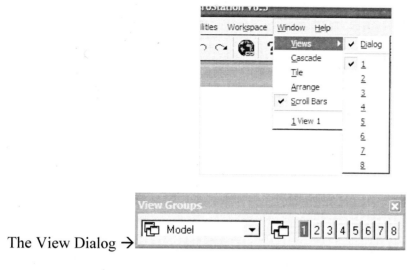

The View Dialog →

When using the View Dialog, simply press and depress the corresponding view number to turn it on and off.

Since more than one view window can be open at one time, the drawing area may become cluttered. To alleviate this problem, Microstation allows for the arrangement of the view windows in three ways:

a. Tiled (views placed side by side with same sizes) – invoked by going to the **Window** pulldown menu> **Tile.**

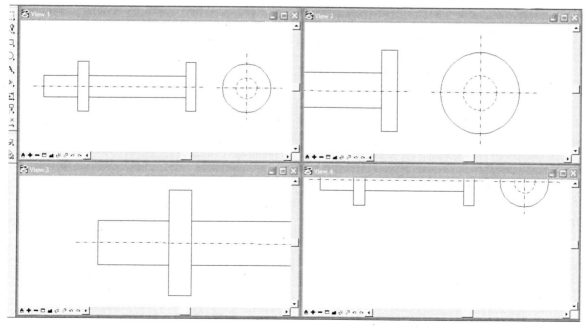

   b. Cascaded (one view placed on top and overlapping the next) - invoked by going to the
      Window pulldown menu> **Cascade.**

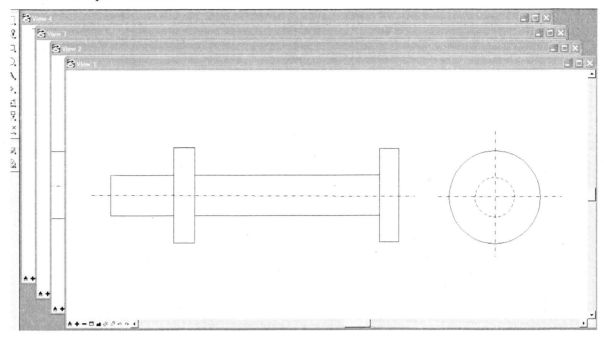

   c. Arranged (randomly placed beside each other with varying sizes) - invoked by going to the
      Window pulldown menu> **Arrange.**

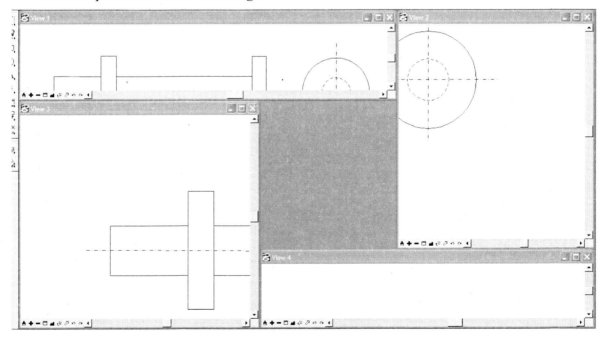

NOTE: _Windows are sized and moved to fit all open views in the application window, as close
to its original size and location as possible._

Using any of the aforementioned arrangements stated above automatically changes the size of
each view window. However, there are other means to sizing and closing a view window,
such as the following methods popularly used in Microsoft Windows programs:

1. Dragging the border of the view window at any location, or at its corners.
2. Minimize the view window by pressing the dash/underscore button, left of the 3 buttons located at the upper right corner of the view window. This temporarily hides the view of the window to the bottom of the application window.
3. Maximize the view window (fully fit the drawing area) by pressing the square button, middle of the 3 buttons located at the upper right corner of the view window. This increases the view of the window to the maximum size allowed to fit in the application window.

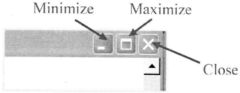

II. View Control

View Control allows for the manipulation and selection of certain areas of the design/drawing to be displayed or viewed. The view controls are located at the lower left corner of each view window.

a. The view control options (shown above from left to right):

1. Update ⌗
   - The view's display will be regenerated and refreshed.

2. Zoom In ✚
   - The view's display will be zoomed IN by the specified zoom ratio (entered in the Tool Setting Box), when a reference point is picked in the design plane. The point picked represent the center of the resulting zoomed in display area.

3. Zoom Out ▬
   - The view's display will be zoomed OUT by the specified zoom ratio (entered in the Tool Setting Box), when a reference point is picked in the design plane. The point picked represent the center of the resulting zoomed out display area.

4. Window Area ▭
   - The view's display will be zoomed IN at a SPECIFIED location in the design plane based upon two reference points that was picked. The points picked represent the size (opposite corners) of the resulting window display area.

5. Fit View ◢
   - All contents in the display will be fitted in the view window, based upon the option chosen in the Tool Setting box.
     The options are:

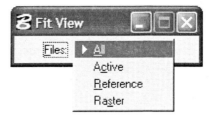

- All – all graphics type embedded into the drawing
- Active – the object derived in and specific to the drawing
- Reference – reference file imported into the drawing
- Raster files – raster image (bmp, jpg, etc.) imported into the drawing

6. Rotate View
   - The view's display will be rotated within the view window or reverted back to original orientation, based upon the option chosen in the Tool Setting box. The options are:

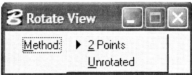

   - 2 points - Two points are picked to initiate the rotation. The 1st point is the center point of which to rotate about, and the 2nd point will define a point/location on the x-axis of the resulting rotated orientation. NOTE: *This option functions differently in 3D.*
   - Unrotated - It rotates the view to its original orientation, considered reversal of a rotation.
   - It will appear as if the drawing is rotated.

7. Pan
   - The view's display will shift/move per the movement of the mouse, based upon the option chosen in the Tool Setting box. The options are:

   - Dynamic Display - The view window will follow or not follow the cursor as it is moved, based upon whether the checkbox option of Dynamic Display option is checked (ON) or unchecked (OFF), respectively.
   - Two points are picked to initiate the move. The 1st point is where to move the view from, and the 2nd point is where to move the view to.

8. View Previous
   - It displays the last displayed view orientation.

9. View Next
   - It undoes what View Previous did. It displays the view before the "View Previous" operation was initiated.

b. Scrolling

Scrolling works like pan, except that sliders (scroll bars), located at the right and bottom of the view window, control the scrolling/shifting of the view up, down, left, and right, respectively.

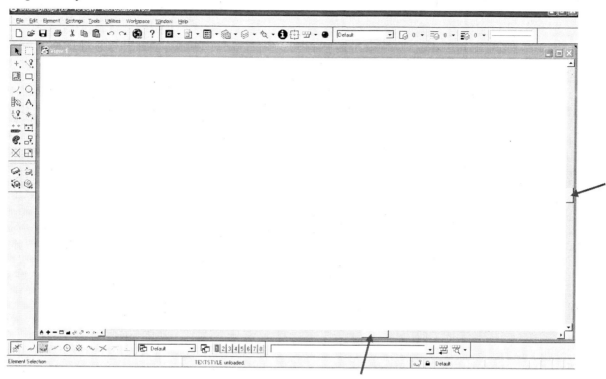

The scroll bars can be turned on and off by placing a check mark beside the Scroll Bars option in the **Window** pulldown menu.

III. Setting View Attributes

View Attributes are features that are displayable in a view. The ability to control the visibility of these features are accessed by going to the **Settings** pulldown menu> **View Attribute** or pressing **Ctrl** and **B** on the keyboard.

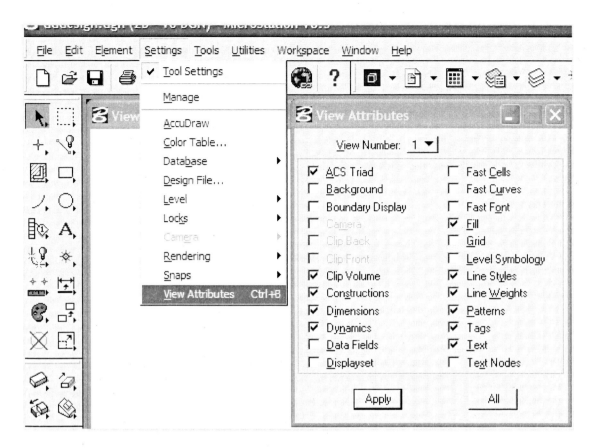

The View Attributes window controls whether elements and/or drawing aids are shown/displayed in one or all view windows. Some attributes are:
- Grid
- Text
- Dimensions

If a box beside a option is checked or unchecked, that particular attribute WILL or WILL NOT be shown, respectively, in the active or all view window(s). They can be applied:
- to the active view by pressing the **Apply** button
- to all views by pressing the **All** button

Chapter 3: Elements

In science, elements are the building blocks of things used and/or made by man. Aluminum (identified as Al in the Periodic Table of Elements) is a metal commonly used in various industries because of its excellent strength-vs-weight mechanical properties. In Microstation, element is the name used for objects created by the program. These objects are the derivatives or building blocks of a completed design. They are geometric shapes (circles, lines, arcs, and etc.) that are combined together to make a part. The parts are assembled together to make the goods, tools, equipment, and machines that are used in everyday life. After completing this chapter, you'll be able to create the following elements:

- Lines
- Blocks
- Polygons
- Circles
- Ellipses
- Arcs
- Points

I. Element/Object Construction and Placement

Drawing is a simple process of using the mouse to pick points on the screen that result in an element being drawn. Most drawing tools allows for the creation of elements that are either linear or circular. The typical drawing tools are located in the Main Palette, which is accessed by going to the **Tools** pulldown menu> **Main**> **Main**.

a. Linear Elements

There are many line tools available in Microstation, however, only some will be covered and discussed. Line tools are accessed from the Linear Elements sub-palette/toolbox in the Main Palette.

NOTE: *When creating lines to and from an existing object, starting and end points are picked by using snaps, to accurately pick a location on the existing object.*

i. Line

A line can be placed with/without a specific length and/or specific angle.

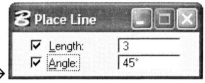

1. The Tool Settings box will look like this →
2. Operating Procedure:
   - Input Available Options:
     - "Length" – input size for each line segment
     - "Angle" – input angle for each line segment

     NOTE: *press <Enter> after each data input to lock in the number for use. If boxes are checked 1st, the data input will be the default line length and angle.*
   - Pick a point on the screen where the line will start from.
   - If "Length" and/or "Angle" is not inputted, pick a point where the line will end.

   <u>Example:</u> Line created per length of 3" and angle of 45°.

NOTE: *The line command will always be active until the reset button (right mouse button) is picked. This allows for continuously drawing.*

ii. Smartline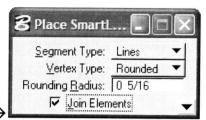

A smartline is in essence an intelligent or smart tool capable of creating more than just lines. It can create individual segments (lines or arcs), or it can create strings of segments joined together at the ends. It can also have sharp, rounded or chamfered vertices between the segments.

1. The Tool Settings box will look like this →
2. Operating Procedure:
    ▪ Choose and/or Input Available Options:
        - "Segment Type" – choose each line segment style
            o Lines
            o Arcs
        - "Vertex Type" – choose vertex style between each line segment
            o Sharp
            o Rounded
            o Chamfered
        - "Rounding Radius/Chamfer Offset" – input the size, if Vertex Type is set to Rounded or Chamfered, respectively.
        - "Join Elements" – place a checkmark beside the feature so the element(s) will be joined at the ends
    ▪ Pick a point where the line will start from.
    ▪ Pick another point to create the first line segment or continue picking to create addition line segments.
    ▪ Reset (right mouse button).

Example: Smartline created per "Lines" method and "Rounded" Vertex Type (5/16 radius).

b. Polygons

In Microstation, a polygon fall under the category of one single enclosed shape made up of multiple line segments. When manipulated, the whole element will be affected, not each line. Placing of these shapes can be accomplished by choosing a tool from the Polygons sub-palette/toolbox in the Main Palette.

i. Blocks ▯

A block is a four-sided enclosed shape made up of two sets of adjoined parallel and perpendicular sides, like a rectangle or square. It can be placed orthogonal or rotated with respect to the X and Y axis of the drawing.

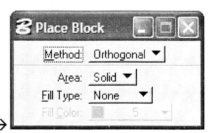

1. The Tool Settings box will look like this →
2. Operating Procedure:
   - Choose Available Options:
     - "Method" – choose the way in which to create the block
       o Orthogonal – each side of the block is parallel to the X and Y axis
       o Rotated – each side of the block is angled to the X and Y axis
     - "Area" – choose the enclosed area of the block
       o Solid *(recommended)*
       o Hole
     - "Fill Type" – choose how to fill/shade the block
       o None – not filled
       o Opaque – filled
       o Outlined – filled with an outline
     - "Fill Color" – choose the color of the filled block
     NOTE: *If "Fill Type" is "Opaque", "Fill Color" is the same as the Active Color used in the Active Attributes toolbox. However, if "Fill Type" is "Outlined", then the outline color is the same as the Active Color used in the Active Attributes toolbox, and "Fill Color" is a separate color.*
   - Pick a point to start the block, which becomes a corner of the block.
   - (For "Orthogonal" method ONLY) – pick another point to complete the block, which is the opposite corner of the block (diagonal from the first/start point).
   - (For "Rotated" method ONLY) – pick another point at any location, which will be the other corner of the same side of the block, and determines the orientation/angle of the block.
   - (For "Rotated" method ONLY) – pick another point to complete the block, which is at the diagonal from the first/start point (the opposite corner of the block).

Example: Block created per:

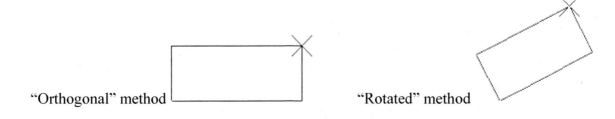

"Orthogonal" method   "Rotated" method

ii. Shape ⊿

A shape is simply an enclosed object made up of adjoined line segments, each at their own length and angle. It is basically a combination of the Smartline and Place Line tool, with the addition of each segment being joined and the final object being closed.

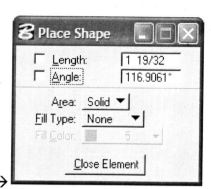

1. The Tool Settings box will look like this →
2. Operating Procedure:
    ▪ Choose and/or Input Available Options:
        - "Length" – input size for each line segment in the shape
        - "Angle" – input angle for each line segment in the shape
        NOTE: *press <Enter> after each data input to lock in the number for use. If boxes are checked 1ˢᵗ, the data input will be the default line segment length and angle.*
        - "Area" – choose the enclosed area of the shape
            o Solid *(recommended)*
            o Hole
        - "Fill Type" – choose how to fill/shade the shape
            o None – not filled
            o Opaque – filled
            o Outlined – filled with an outline
        - "Fill Color" – choose the color of the filled shape
        NOTE: *If "Fill Type" is "Opaque", "Fill Color" is the same as the Active Color used in the Active Attributes toolbox. However, if "Fill Type" is "Outlined", then the outline color is the same as the Active Color used in the Active Attributes toolbox, and "Fill Color" is a separate color.*
        - "Close Element" – pick to automatically close/finish the element
    ▪ Pick the start point of the shape.
    ▪ Pick the next point (with/without input of the Length and/or Angle of the line segment).
    ▪ Repeat the previous step until the final object is created, or by picking "Close Element" in the Tool Setting box.

Example: Shape created.

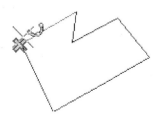

iii. Orthogonal Shape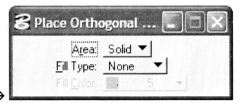
An orthogonal shape is an enclosed object made up of adjoined line segments, where each line segment is perpendicular to the next, and every other line segment parallel to each other. It is a combination of the Block and Shape tool.

1. The Tool Settings box will look like this →
2. Operating Procedure:
    ▪ Choose Available Options:
        - "Area" – choose the enclosed area of the orthogonal shape
            o Solid *(recommended)*
            o Hole
        - "Fill Type" – choose how to fill/shade the orthogonal shape
            o None – not filled
            o Opaque – filled
            o Outlined – filled with an outline
        - "Fill Color" – choose the color of the filled orthogonal shape
        NOTE: *If "Fill Type" is "Opaque", "Fill Color" is the same as the Active Color used in the Active Attributes toolbox. However, if "Fill Type" is "Outlined", then the outline color is the same as the Active Color used in the Active Attributes toolbox, and "Fill Color" is a separate color.*
    ▪ Pick the start point of the shape.
    ▪ Pick the next and consecutive points until the final object is created.
    NOTE: *The shape is completely drawn when the 1ˢᵗ point is picked again as the last point.*
Example: Orthogonal Shape created (with "Opaque" fill type).

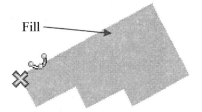
Fill

iv. Regular Polygons
A regular polygon is an enclosed shape made up of adjoined line segments of equal length. The number of sides range from 3 to 4999. The size of the polygon is determined by it being inscribed/circumscribed about an imaginary circle of a certain radius, or by one edge of the polygon.

1. The Tool Settings box will look like this →
2. Operating Procedure:
   - Choose and/or Input Available Options:
     - "Method" – choose the way in which to create the polygon
       o Inscribed – polygon is placed inside an imaginary circle
       o Circumscribed – polygon is placed outside an imaginary circle
       o By Edge – polygon is placed per two points, which determines the length of the sides
     - "Edges" – input the number of sides for the polygon
     - "Radius" – input the radius size of the imaginary circle used to create the polygon (Inscribed/Circumscribed method ONLY)
     - "Area" – choose the enclosed area of the polygon
       o Solid *(recommended)*
       o Hole
     - "Fill Type" – choose how to fill/shade the polygon
       o None – not filled
       o Opaque – filled
       o Outlined – filled with an outline
     - "Fill Color" – choose the color of the filled polygon
     NOTE: *If "Fill Type" is "Opaque", "Fill Color" is the same as the Active Color used in the Active Attributes toolbox. However, if "Fill Type" is "Outlined", then the outline color is the same as the Active Color used in the Active Attributes toolbox, and "Fill Color" is a separate color.*
   - For "Inscribed/Circumscribed" methods ONLY) – pick a point where the center of the polygon will be
   - (For "Inscribed/Circumscribed" methods ONLY) – pick another point where one of the vertices of the polygon will be *(For an Inscribed polygon, this point will basically lie on the circumference of the imaginary circle).*
   - (For "By Edge" method ONLY) – pick a point where one of the vertices of the polygon will be
   - (For "By Edge" method ONLY) – pick a point where another vertex of the polygon will be, which determines the orientation/angle of the polygon and length of each side of the polygon (therefore, the polygon's overall size).

Example: Regular Polygon created per "Inscribed" method (6 edges, with "Outlined" fill type).

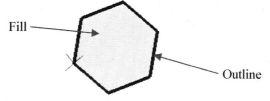

c. Circular Elements

Circles, Arcs, and Ellipses are the usual suspects when considering circular objects, as they sometime share common characteristics. Circles have a diameter and radius, while Arcs have a radius only. Ellipses have neither, but have 2-axes, primary and secondary, which are perpendicular to each other. Nonetheless, Microstation groups Circle and Ellipse tools in the Ellipses sub-palette/toolbox, and places the Arc tool in the Arcs sub-palette/toolbox, both which are found in the Main Palette.

i. Circles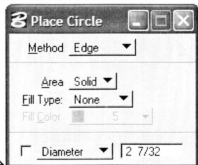

A circle is a perfectly symmetrical, enclosed, curved shape that can be created based off its center point location, diameter/radius, or points on its circumference.

1. The Tool Settings box will look like this →
2. Operating Procedure:
   ▪ Choose and/or Input Available Options:
      - "Method" – choose the way in which to create the circle
         ○ Center
         ○ Edge
         ○ Diameter
      - "Area" – choose the enclosed area of the circle
         ○ Solid *(recommended)*
         ○ Hole
      - "Fill Type" – choose how to fill/shade the circle
         ○ None – not filled
         ○ Opaque – filled
         ○ Outlined – filled with an outline
      - "Fill Color" – choose the color of the filled circle

      NOTE: *If "Fill Type" is "Opaque", "Fill Color" is the same as the Active Color used in the Active Attributes toolbox. However, if "Fill Type" is "Outlined", then the outline color is the same as the Active Color used in the Active Attributes toolbox, and "Fill Color" is a separate color.*

      - "Diameter/Radius" – choose diameter or radius, and input a known corresponding size. *Press <Enter> after the data input to lock in the number for use. If box is checked 1ˢᵗ, the data input will be the default circle diameter or radius.*

- (For "Center" method ONLY) – pick a point that will represent the center for the circle. *If the diameter or radius is inputted, all that is required is to pick a point where the "center" of the circle will be, and the operating procedure ends here.*

- (For "Center" method ONLY) – pick another point that will represent a location on the circumference of the circle, which will complete the design of the circle.

- (For "Diameter" method ONLY) – pick a point that will represent the 1st point of the diameter for the circle.

- (For "Diameter" method ONLY) – pick another point, which is the opposite/ second end of the diameter for the circle.

- (For "Edge" method ONLY) – pick a point that will represent a location on the circumference of the circle.

- (For "Edge" method ONLY) – pick two other points that will represent other locations on the circumference of the circle, which will complete the design of the circle. *If the diameter or radius is inputted, all that is required is to pick one other point, not two.*

<u>Example:</u> Circle created per "Edge" method *(features shown for reference).*

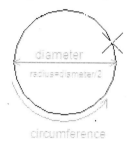

ii. Ellipses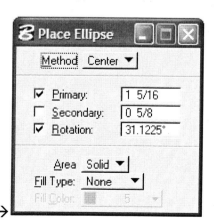

An ellipse is a symmetrical, enclosed, curved shape that can be created based off its center point location, two (2) axis size, or points on its edge.

1. The Tool Settings box will look like this →

2. Operating Procedure:
   - Choose and/or Input Available Options:
     - "Method" – choose the way in which to create the ellipse
       o Center
       o Edge
     - "Area" – choose the enclosed area of the ellipse
       o Solid *(recommended)*
       o Hole
     - "Fill Type" – choose how to fill/shade the ellipse
       o None – not filled
       o Opaque – filled
       o Outlined – filled with an outline
     - "Fill Color" – choose the color of the filled ellipse

     <u>NOTE:</u> *If "Fill Type" is "Opaque", "Fill Color" is the same as the Active Color used in the Active Attributes toolbox. However, if "Fill Type" is "Outlined", then the outline color is the same as the Active Color used in the Active Attributes toolbox, and "Fill Color" is a separate color.*

     - "Primary" – input a known size for the primary axis of the ellipse. *If all of the optional data is inputted, all that is required is to pick a point where the "center" for the ellipse will be.*
     - "Secondary" – input a known size for the secondary axis of the ellipse. *If all of the optional data is inputted, all that is required is to pick a point where the "center" for the ellipse will be.*
     - "Rotation" – input a known angle for the ellipse. *If all of the optional data is inputted, all that is required is to pick a point where the "center" for the ellipse will be.*
   - (For "Center" method ONLY) – pick a point that will represent the center for the ellipse. *If all optional data is inputted, all that is required is to pick a point where the "center" of the ellipse will be, and the operating procedure ends here.*
   - (For "Center" method ONLY) – pick another point that will represent a location on the edge of the ellipse, which forms the primary axes and orientation/rotation for the ellipse.
   - (For "Center" method ONLY) – pick another point that will represent another location on the edge of the ellipse, which forms the secondary axes for the ellipse. This will complete the design of the ellipse.
   - (For "Edge" method ONLY) – pick a point that will represent a location on the edge of the ellipse. *If all optional data is inputted, all that is required is to pick a point where the "edge" of the ellipse will be, and the operating procedure ends here.*

■ (For "Edge" method ONLY) – pick two other point that will represent other locations on the edge of the ellipse, which will complete the design of the ellipse.

<u>Example:</u> Ellipse created per "Center" method *(features shown for reference).*

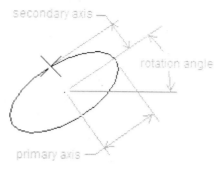

iii. Arcs

An arc is an open curved shape, looking like part of a circle, and can be created based off its center point location, radius, or points on its edge.

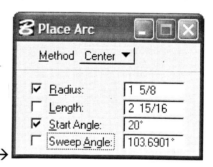

1. The Tool Settings box will look like this →
2. Operating Procedure:
   ■ Choose and/or Input Available Options:
     - "Method" – choose the way in which to create the circle
       o Center
       o Edge
     - "Radius" –input a known size
     - "Length" –input a known size
     - "Start Angle" –input a known angle to start the arc from
     - "Sweep Angle" –input a known angle that the arc will span
     NOTE: *press <Enter> after each data input to lock in the number for use. The length of the arc is equivalent to the difference between the start angle and sweep angle of the arc. Therefore, a known start angle and sweep angle can replace the length of the arc. Secondly, a known length of the arc CANNOT be inputted with both a known start angle and a known sweep angle, EITHER the start angle OR sweep angle has to be inputted. If boxes are checked 1ˢᵗ, the data input will be the default arc settings.*

- (For "Center" method ONLY) – pick a point that will represent a location on the edge and start of the arc. *If any combination of three of the logically, optional data is inputted, all that is required is to pick a point where the "center" for the arc will be, and the operating procedure ends here.*
- (For "Center" method ONLY) – pick another point that will represent the center point for the arc.
- (For "Center" method ONLY) – pick another point that will represent the end of the arc, which will complete the design of the arc.
- (For "Edge" method ONLY) – pick a point that will represent a location on the edge and start of the arc. *If any combination of three of the logically, optional data is inputted, all that is required is to pick a point where an "endpoint" for the arc will be, and the operating procedure ends here.*
- (For "Edge" method ONLY) – pick two other points that will represent other locations on the edge of the arc, which will complete the design of the arc.

Example: Arc created per "Center" method *(features shown for reference).*

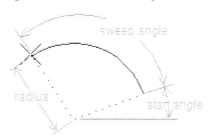

d. Points

In Microstation, a point is a single entity that can be either represented as an element/dot (typical), character, or cell (similar to blocks in AutoCAD; *see Microstation v8: Simplified, Vol. 1, Part B*). These are known as the "Point Type". The point tools are accessible via the Points sub-palette/toolbox, found in the Main Palette:

NOTE: *If a point is represented as an Element, then the size of the point is based off the line weight. Character size is based off Text Style height. Cell size is based off the Active Scale.*

Points can be placed in the drawing in various ways, dependent on the Point tool chosen:

i. Place Point ✛
Places a point anywhere picked with the mouse

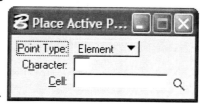

1. The Tool Settings box will look like this →
2. Operating Procedure:
    ▪ Choose and/or Input Available Options:
        - "Point Type" – choose the style for the point
            o Element
            o Character
            o Cell
    NOTE: *if "Character" or "Cell" is chosen as the Point Type to use for the point, input a character (from keyboard) or cell name, respectively.*
    ▪ Pick the location where a point will be placed.
    ▪ Continue picking points, or Reset (right mouse button) to end the process.
<u>Example:</u> A point created.

■

ii. Place Point between Data Points
Places a set of equally spaced points between two specific points/locations

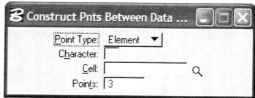

1. The Tool Settings box will look like this →
2. Operating Procedure:
    ▪ Choose and/or Input Available Options:
        - "Point Type" – choose the style for the point
            o Element
            o Character
            o Cell
    NOTE: *if "Character" or "Cell" is chosen as the Point Type to use for the point, input a character (from keyboard) or cell name, respectively.*
        - "Points" – input number of points to be placed
    ▪ Pick the location where the 1st point will be placed.
    ▪ Pick a second location where the last point will be placed.
    ▪ Continue picking points, or Reset (right mouse button) to end the process.
<u>Example:</u> 3 points created between two locations.

■

■

■

iii. Place Point ONTO Element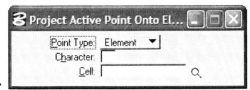
      Places a point anywhere on an element

1. The Tool Settings box will look like this →
2. Operating Procedure:
      ▪ Choose and/or Input Available Options:
          - "Point Type" – choose the style for the point
              o Element
              o Character
              o Cell
        NOTE: *if "Character" or "Cell" is chosen as the Point Type to use for the point,*
                    input a character (from keyboard) or cell name, respectively.
      ▪ Pick a location on the element where a point will be placed.
<u>Example:</u> 1 point created on an arc.

iv. Place Point at Intersection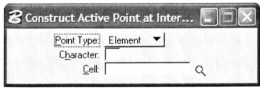
      Places a point at the intersection of two elements

1. The Tool Settings box will look like this →
2. Operating Procedure:
      ▪ Choose and/or Input Available Options:
          - "Point Type" – choose the style for the point
              o Element
              o Character
              o Cell
        NOTE: *if "Character" or "Cell" is chosen as the Point Type to use for the point,*
                    input a character (from keyboard) or cell name, respectively.
      ▪ Pick the 1st element where the point will intersect with.
      ▪ Pick the 2nd element where the point will intersect with.
<u>Example:</u> 1 point created at intersection of 2 lines.

v. Place Point ALONG element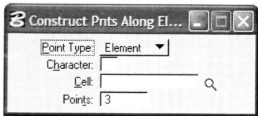
    Places a set of equally spaced points along an element, between two specific
    points/locations on the element

1. The Tool Settings box will look like this →
2. Operating Procedure:
    ▪ Choose and/or Input Available Options:
        - "Point Type" – choose the style for the point
            o Element
            o Character
            o Cell
    NOTE: *if "Character" or "Cell" is chosen as the Point Type to use for the point,
        input a character (from keyboard) or cell name, respectively.*
        - "Points" – input number of points to be placed
    ▪ Pick the location on the element where the 1st point will be placed.
    ▪ Pick a second location on the element where the last point will be placed.
    NOTE: *This tool functions as a combination of the Place Point between Data
        Points tool and Place Point ONTO Element tool.*

Example: 3 points created on an arc.

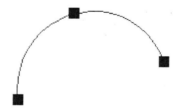

vi. Place Point @ DISTANCE along element
     Places a point at a given distance from a data point/location that is picked

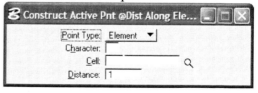

1. The Tool Settings box will look like this →
2. Operating Procedure:
    ▪ Choose and/or Input Available Options:
        - "Point Type" – choose the style for the point
            o Element
            o Character
            o Cell
    NOTE: *if "Character" or "Cell" is chosen as the Point Type to use for the point,
        input a character (from keyboard) or cell name, respectively.*

- Input the distance in which the point will be placed from a reference location
- Pick the location on the element where the point will be referenced from.
- Accept (left mouse button) or Pick a second location on the element to place the point.

Example: 1 point created at a distance (1") from a reference point on the arc.

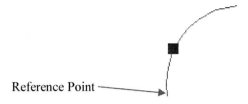

Reference Point

e. Delete

After the elements have been drawn one needs to be prepared to erase it if drawn incorrectly. Deleting or erasing is accomplished with the delete tool, which is found in the Main Palette

1. Operating Procedure:
   - Pick the element
   - Reset (right mouse button) to finish the command
   - or pick another element (as many as needed)

NOTE: *more than one element can be deleted, if the Element Selection tool highlights the elements to make it active.*

Chapter 4: Manipulate

Everything in life is not guaranteed. There is always room for adjustments and/or changes. When a product is made for the first time, the result may not completely be what is expected. Some aspect of the product may need to be manipulated or modified. Manipulation is an adjustment that tends to relocate, re-orient, and/or resize the object. Modification will be discussed later. After completing this chapter, you'll be able to demonstrate the following manipulation techniques:

- Copy
- Move
- Move/Copy Parallel
- Scale
- Rotate
- Mirror
- Align Edges
- Construct Array

I. Manipulate Elements

Manipulating an element is simply adjusting some existing aspect of the element. This adjustment is accomplished by actions such as moving, copying, rotating, or scaling the element, amongst other functions. The Manipulate sub-palette/toolbox is accessed from the Main Palette.

a. Copy *Element*
    This tool allows for the copy/duplication of an element, and its positioning, at any location.

The Tool Settings box will look like this →
  i. Operating Procedure:
      1. Choose and/or Input Available Options:
          ▪ "Copies" – input the number of copies needed, and press <Enter> to place a checkmark beside that feature to accept that value
          ▪ "Use Fence" – place a checkmark beside the feature (and choose the fence mode from the associated pulldown menu) to use a fence to copy specific or multiple elements
      2. Pick the element(s) to copy.
      3. Pick the location for the newly copied element(s). *If more than one copy is inputted, the distance and orientation between the original and the first copy will be the interval between all consecutive copies.*
      NOTE: *more than one element can also be manipulated if the Element Selection tool highlights the elements to make it active.*

b. Move *Element*
    This tool allows for the movement of an element to any location.

The Tool Settings box will look like this →
  i. Operating Procedure:
      1. Choose and/or Input Available Options:
          ▪ "Copies" - place a checkmark beside the feature if copies are needed. *If checked, the tool switches to "Copy Element" and function as such.*
          ▪ "Use Fence" – place a checkmark beside the feature (and choose the fence mode from the associated pulldown menu) to use a fence to move specific or multiple elements
      2. Pick the element(s) to move.
      3. Pick the element(s) new location.
      NOTE: *more than one element can also be manipulated if the Element Selection tool highlights the elements to make it active.*

c. Move/Copy Parallel

This tool allows for the offsetting of an element (like in AutoCAD), to a certain distance from the specified element. The element is moved/copied parallel to the original location.

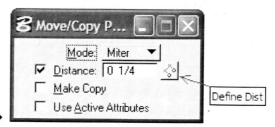

The Tool Settings box will look like this →

i. Operating Procedure:
1. Choose and/or Input Available Options:
   - "Mode" – choose the way in which to move/copy parallel the element
     - Miter *(recommended)* – the offset object will have sharp vertices/corners, and tries to keep all offset lines parallel to the original, but will cut off excess if can't
     - Round – the offset object will have rounded vertices/corners
     - Original – the offset object will have sharp vertices/corners, and tries to keep all offset lines parallel to the original
   - "Distance" - input an offset distance, and press <Enter> to place a checkmark beside that feature to accept that value
   - "Make Copy" - place a checkmark beside the feature if a copy is needed. *If checked, the originally picked element(s) will stay in its original location and a copy of that element(s) will be positioned per the offset distance.*
   - "Use Active Attributes" – place a checkmark beside the feature to allow the offset element to have the characteristics of the active (current) attributes, like level, color, style, and weight
2. Pick the element to offset.
3. Pick the side where the offset element will be located.
4. Accept (left mouse button).
5. Reset (right mouse button).

NOTE: *if the offset distance is not known, but can be derived from an existing distance in the drawing, use the "Define Dist" button to pick two points in the drawing.*

Example: element offset per "Round" method (offset distance of ¼).

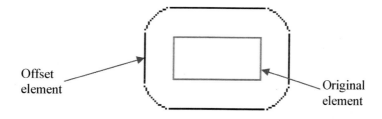

d. Scale

This tool allows for the scaling (increase or decrease size) of an element.

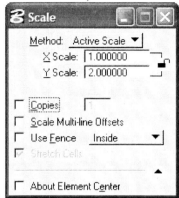

The Tool Settings box will look like this →

i. Operating Procedure:
1. Choose and/or Input Available Options:
     ▪ "Method" – choose the way in which to scale the element
          - Active Scale – uses a specific angle to scale the element(s)
          - 3 points – uses three points to scale the element(s)
     ▪ "Copies" - input the number of copies needed, and press <Enter> to place a
          checkmark beside that feature to accept that value. *If checked, the
          originally picked element(s) will stay in its original location and a
          copy of that element(s) will be scaled.*
     ▪ "Use Fence" – place a checkmark beside the feature (and choose the fence mode
          from the associated pulldown menu) to use a fence to scale
          specific or multiple elements
     ▪ "Scale Multi-line Offsets" - place a checkmark beside the feature to allow the
          offsets of the lines in a multi-line to be scaled.
     ▪ "About Element Center" - place a checkmark beside the feature so the
          element(s) will be scaled about the center of the
          element
2. Pick the element(s) to scale.
3. Pick the origin location, which is the point where the element will be scaled from.
4. (For "3 Points" method ONLY) - pick any location in drawing to use as a control
          point, like an handle, to manually control the
          scaling of the element(s).
5. Accept (left mouse button). *If more than one copy is inputted, the scale factor
     determined between the original and the first copy will be that use for all
     consecutive copies.*
NOTE: *more than one element can also be manipulated if the Element Selection tool
          highlights the elements to make it active.*
Example: element scaled per "Active Scale" method (X and Y scale factors of 0.5).

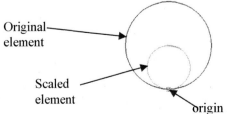

Original—
element

Scaled
element

origin

e. Rotate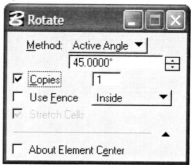

This tool allows for the rotation of an element, about an axis.

The Tool Settings box will look like this →

i. Operating Procedure:
  1. Choose and/or Input Available Options:
      ▪ "Method" - choose the way in which to rotate the element
          - Active Angle - uses a specific angle to rotate the element(s)
          - 2 points – uses two points to rotate the element(s)
          - 3 points – uses three points to rotate the element(s)
      ▪ "Copies" - input the number of copies needed, and press <Enter> to place a
          checkmark beside that feature to accept that value. *If checked, the
          originally picked element will stay in its original location and a copy
          of that element will be rotated*
      ▪ "Use Fence" – place a checkmark beside the feature (and choose the fence mode
          from the associated pulldown menu) to use a fence to rotate
          specific or multiple elements
      ▪ "About Element Center" - place a checkmark beside the feature so the
          element(s) will be rotated about the center of the
          element
  2. Pick the element(s) to rotate.
  3. Pick a location for the rotation axis (pivot point).
  4. (For "3 Points" method ONLY) - pick a location in the drawing to use as a control
          point, like a handle, to manually control the
          rotation of the element(s).
  5. (For "2 Points" and "3 Points" methods ONLY) - move the pointer and manually
          control the rotation of the
          element(s).
  6. (For "2 Points" and "3 Points" methods ONLY) - pick a location in the drawing to
          complete the rotation process.
  7. Reset (right mouse button). *If more than one copy is inputted, the rotation angle
      determined between the original and the first copy will be that use for all
      consecutive copies.*
  <u>**NOTE:**</u> *more than one element can also be manipulated if the Element Selection tool
      highlights the elements to make it active.*
  <u>Example:</u> element rotated per "Active Angle" method (45°).

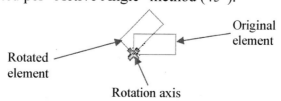

Original
element

Rotated
element

Rotation axis

f. Mirror
   This tool allows for the mirroring of an element, about a imaginary line or two picked points.

The Tool Settings box will look like this →

   i. Operating Procedure:
       1. Choose Available Options:
           ▪ "Mirror About" – choose the way in which to mirror the element
               - Horizontal – mirrors the element about an imaginary horizontal line
               - Vertical – mirrors the element about an imaginary vertical line
               - Line – mirrors the element about 2 picked points (creates an imaginary line)
           ▪ "Make Copy" - place a checkmark beside the feature if a copy is needed. *If checked, the originally picked element will stay in its original location and a copy of that element will be positioned per the "Mirror About" method chosen.*
           ▪ "Mirror Text" - place a checkmark beside the feature to have text mirrored.
           ▪ "Mirror Multi-line Offset" - place a checkmark beside the feature to allow the offsets of the lines in a multi-line to be mirrored
           ▪ "Use Fence" – place a checkmark beside the feature (and choose the fence mode from the associated pulldown menu) to use a fence to mirror specific or multiple elements
       2. Pick the element(s) to mirror.
       3. Pick the location of the new element, based upon the "Mirror About" method.
       NOTE: *more than one element can also be manipulated if the Element Selection tool highlights the elements to make it active.*

       Example: element mirrored per "Line" method.

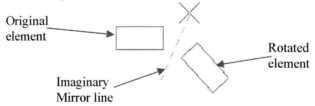

g. Align *Edges*
   This tool allows for the aligning of an element with another specific element.

The Tool Settings box will look like this →

i. Operating Procedure:
1. Choose Available Options:
- "Align" – choose the way in which to align the elements
  - Top – object(s) are aligned to the top of another
  - Bottom – object(s) are aligned to the bottom of another
  - Left – object(s) are aligned to the left of another
  - Right – object(s) are aligned to the right of another
  - Horiz Center – object(s) are aligned to the horizontal center of another
  - Vert Center – object(s) are aligned to the vertical center of another
  - Both Center – object(s) are aligned to center (horizontal and vertical) of another
- "Use Fence" – place a checkmark beside the feature (and choose the fence mode from the associated pulldown menu) to use a fence to align specific or multiple elements
2. Pick the element (base element) in which other element(s) will be "aligned to".
3. Pick the element to be "aligned".
4. Accept (left mouse button).

NOTE: more than one element can also be manipulated if the Element Selection tool highlights the elements to make it active.

Example: element aligned per "Right" method.

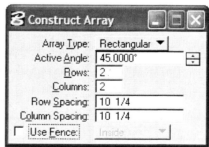

h. *Construct* Array

This tool allows for the array (put in a matrix) of an element, either rectangular or radial (in a circle manner).

The Tool Settings box will look like this →

i. Operating Procedure:
1. Choose and/or Input Available Options:
- "Array Type" – choose the way in which to array the element
  - Rectangular – array is created in a rectangular manner, per rows and columns
  - Polar – array is created in a circular manner
  - *NOTE: For Polar Array Type, the result of the array is based off:*
    - *Total Rotating (Sweep) Angle - the span of the array*
    - *Delta Angle - angle between the items in the array*
    - *Total number of Items in the Array*

For Total Rotating (Sweep) Angles EQUAL to 360 degrees,
Delta Angle = Total Rotating (Sweep) Angle
(Total number of Items in the Array)

For Total Rotating (Sweep) Angles LESS THAN 360 degrees,
Delta Angle = Total Rotating (Sweep) Angle
(Total number of Items in the Array - 1)

- "Active Angle" (Rectangular Arrays) – input angle in which to orient the array
- "Row" (Rectangular Arrays) – input number of rows in the array
- "Column" (Rectangular Arrays) – input number of columns in the array
- "Row Spacing" (Rectangular Arrays) – input the spacing size between each row in the array
- "Column Spacing" (Rectangular Arrays) – input the spacing size between each column in the array
- "Items" (Polar Arrays) – input the number of items in the array
- "DeltaAngle" (Polar Arrays) – input the angle between each item in the array
- "Rotate Items" (Polar Arrays) - place a checkmark beside the feature to allow each item in the array to be rotated as it is placed
- "Use Fence" – place a checkmark beside the feature (and choose the fence mode from the associated pulldown menu) to use a fence to array specific or multiple elements

2. Pick the element to use as the basis for the array.
3. (For "Polar" Arrays ONLY) - pick the origin/location where the element array will be based from.

NOTE: *more than one element can also be manipulated if the Element Selection tool highlights the elements to make it active.*

Example: element arrayed per "Polar" method.
Sweep Angle = 180°, and Number of Items = 4, therefore,
DeltaAngle = 180°/(4-1) = 180°/3 = 60°

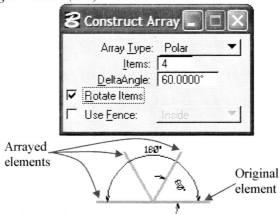

Arrayed elements   180°   Original element

Example: element arrayed per "Rectangular" method (2 rows and 2 columns).

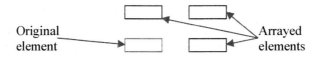

Original element   Arrayed elements

Chapter 5: Modify

Progress is usually uncommon without change. As stated earlier, there is always room for adjustments and/or changes in most things. Adjustments are accomplished through manipulation, while changes are achieved via modification. Modification tends to reshape, subtract from, or add to an existing element/part. After completing this chapter, you'll be able to demonstrate the following modification techniques:

- Modify/Stretch
- Partial Delete
- Extend
- Trim
- Insert/Delete Vertex
- Fillet
- Chamfer

I. Modify Elements

Modifying an element is basically changing some existing feature of the element, affecting the original geometry. The change involves actions such as stretching, extending, or trimming of the element, amongst other functions. The Modify sub-palette/toolbox is accessed from the Main Palette.

a. Modify *Element*

This tool allows for the changing of the geometry/shape of an element. If a circle is picked, it's diameter can be resized via movement of the mouse. If a line/linear object is picked, its endpoint can be repositioned/relocated. If an arc is picked, its endpoint can be repositioned/ relocated, along with changing the radius via movement of the mouse. If a dimension is picked, its text or dimension line can be repositioned or relocated. If a block object is picked, its vertices can be repositioned/relocated and the object skewed (the result depends on the location picked on the block). *There are NO Tool Settings box options available with this tool,* unless a block or enclosed linear shape is picked, where it may look like this:

i. Operating Procedure:

1. Choose and/or Input Available Options: *For enclosed linear shapes/objects ONLY*
   - "Vertex Type" – choose the vertex style between each line segment
     - Sharp
     - Rounded
     - Chamfered
   - "Rounding Radius/ Chamfer Offset" – input the size, if Vertex Type is set to Rounded or Chamfered, respectively
   - "Orthogonal" (Blocks ONLY) – place a checkmark beside the feature to keep the block orthogonal
2. Pick the element to modify.
3. Pick a location in the drawing to change the size/location of the element.

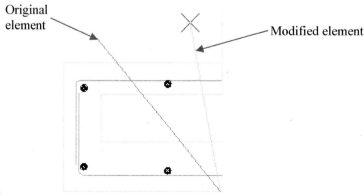

Original element

Modified element

Example: modified line.

b. Partial Delete
This tool allows for the deletion of only part of an element, based upon where is picked. *There are NO Tool Settings box options available with this tool.*
  i. Operating Procedure:
    1. Choose and/or Input Available Options: *NONE*
    2. Pick a location on the element to start the deletion.
    3. Pick another location to end/stop the deletion.
    Example: partially deleted Arc.

c. Extend *Line*
This tool allows for an element to be extended/ retracted to a certain length/distance.

The Tool Settings box will look like this →
  i. Operating Procedure:
    1. Choose and/or Input Available Options:
      ▪ "Distance" – place a checkmark beside the feature and input the distance to extend or retract the line to
      ▪ "From End" - place a checkmark beside the feature so the line can be extended or retracted from the endpoint. ***NOTE:*** *For "Distance" not checked ONLY.*
    2. Pick on the element to extend/retract
    3. Pick a location in the drawing where the linear object will be extended/ retracted to.

d. Extend *Elements* to Intersection
This tool allows for two elements to be extended/ retracted until they intersect with each other, if possible. *There are NO Tool Settings box options available with this tool.*
  i. Operating Procedure:
    1. Choose or Input Available Options: *NONE*
    2. Pick the 1st element to be extended/ retracted.
    3. Pick the 2nd element to be extended/ retracted.
    4. Accept (left mouse button).
    Example: two elements extended to an intersection.

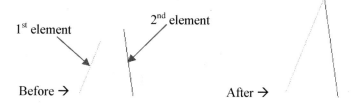

e. Extend *Element* to Intersection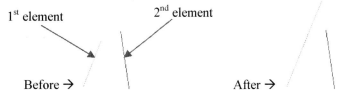
   This tool allows for "one" element to be extended/ retracted until it intersects with another element, if possible. *There are NO Tool Settings box options available with this tool.*
   i. Operating Procedure:
      1. Choose or Input Available Options: *NONE*
      2. Pick the 1st element to be extended/ retracted.
      3. Pick a 2nd element in which to extend/ retract "to".
      4. Accept (left mouse button).
      Example: one element extended to another element where it expected to intersect.

f. Trim *Elements*
   This tool allows for an element to be trimmed, with respect to another element. *There are NO Tool Settings box options available with this tool.*
   i. Operating Procedure:
      1. Choose or Input Available Options: *NONE*
      2. Pick the element (the cutting element) where the trimming will occur at.
      3. Pick the element (the trim element) to be trimmed.
      4. Accept (left mouse button).
      Example: one element trimmed off at the intersection of another.

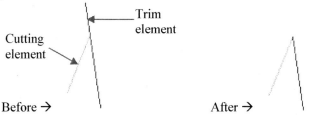

g. Intelli*Trim*
   This tool allows for more than one element to be trimmed, using more than one element as the cutting element (boundary).

   The Tool Settings box will look like this →
   i. Operating Procedure:
      1. Choose Available Options:
         ▪ "Mode" – the way in which trim the elements
            - Quick – requires one cutting element, and multiple elements to be trimmed, extended, or cut
            - Advanced – requires multiple cutting element(s), and multiple elements to be trimmed or extended

- "Operations" – the task in which to do
  - Trim
  - Extend
  - Cut ("Quick" Mode ONLY)

2. Trim ("Quick" Mode)
   - Pick the element (the cutting element) where the trimming will occur at. The element will turn to a hidden line of different color.
   - Draw a temporary line across the element you want to trim.
   - Repeat previous step as many times as needed, or press Reset (right mouse button).

Quick Trim Example:

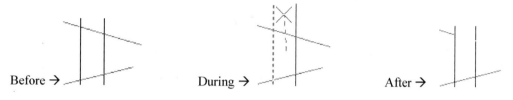

Before →   During →   After →

3. Extend ("Quick" Mode)
   - Pick an element in which to extend "to". The element will turn to a hidden line of different color.
   - Draw a temporary line across the element you want to extend.
   - Repeat previous step as many times as needed, or press Reset (right mouse button) when done.

Quick Extend Example:

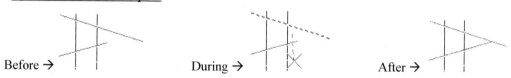

Before →   During →   After →

4. Cut ("Quick" Mode)
   - Draw a temporary line across the element to be cut. This temporary line will act like a saw/scissor, and cut the element.
   - Repeat previous step as many times as needed, or change command.

Quick Cut Example:

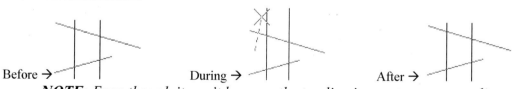

Before →   During →   After →

   NOTE: *Even though it can't be seen, the top line is now two separate lines*

5. Trim ("Advanced" Mode)
   - Pick as many cutting elements (boundary element) as needed. The element will turn to a hidden line of different color.
   - Reset (right mouse button). ***NOTE:*** *The "Select Cutting Elements" button will switch to "Select Element to Trim".*

- Pick as many elements needed to be trimmed. Black dots will be placed at the intersection of the element(s) to be trimmed and the cutting element(s).
- Reset (right mouse button).
- A preview of the trimmed element(s) will be displayed.
- If NOT satisfied with the trimming displayed, choose the alternate trimmed preview by picking on the opposite side of the cutting element.
- If satisfied with the trimming displayed, Reset (right mouse button).

Advanced Trim Example: two elements trimmed between two cutting elements.

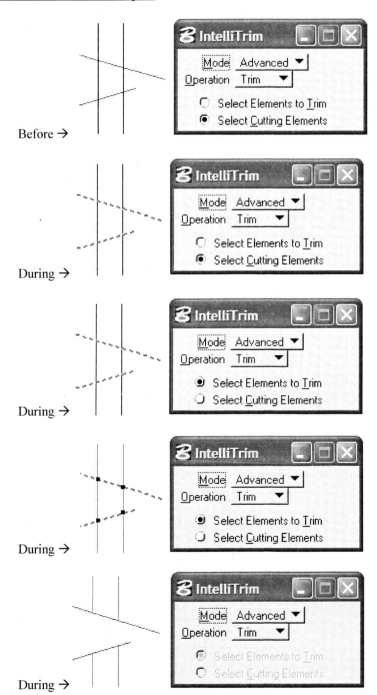

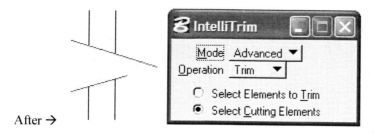

After →

6. Extend ("Advanced" Mode)
   - Pick as many cutting elements (boundary element) you want. The element will turn to a hidden line of different color.
   - Reset (right mouse button). **_NOTE:_** *The "Select Cutting Elements" button will switch to "Select Element to Trim".*
   - Pick as many elements needed to be extended. Black dots will be placed at the intersection of the element(s) to be extended and the cutting element(s).
   - Reset (right mouse button).
   - A preview of the extended element(s) will be displayed.
   - If NOT satisfied with the extension displayed, choose the alternate extension preview by picking on the opposite side of the cutting element.
   - If satisfied with the extension displayed, Reset (right mouse button).

   <u>Advanced Extend Example</u>: two elements extended to a boundary element.

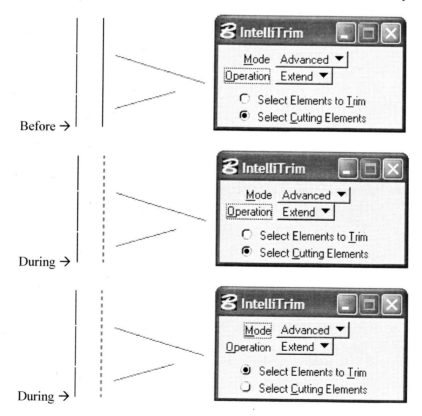

Before →

During →

During →

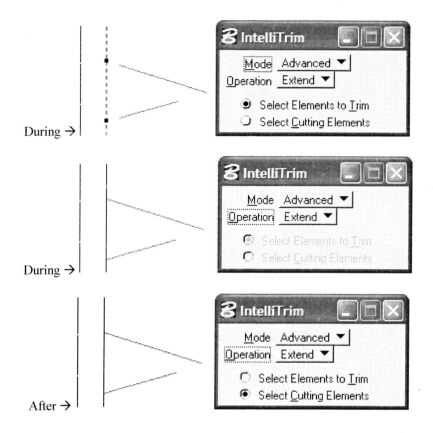

h. *Insert* Vertex

This tool allows for a vertex to be inserted anywhere into an open or enclosed linear element. *There are NO Tool Settings box options available with this tool.*

   i. Operating Procedure:
     1. Choose and/or Input Available Options: *NONE*
     2. Pick the location on the element where a vertex is needed.
     3. Pick more locations or Reset (right mouse button) when done.

   <u>Example:</u>   Before →   After →

i. *Delete* Vertex

This tool allows for a vertex to be deleted from an open or enclosed linear element. When this tool is used on a single line segment, the result will be a point. *There are NO Tool Settings box options available with this tool.*

   i. Operating Procedure:
     1. Choose and/or Input Available Options: *NONE*
     2. Pick the location on the element where you want to delete the vertex.
     3. Pick more locations or Reset (right mouse button) when done.

   <u>Example:</u>   Before →   After →

j. *Construct* Circular Fillet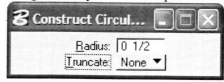
   This tool allows for a fillet (radius) to be created between two picked elements. If the radius of the fillet is bigger than the length of any of the two picked elements, then an "illegal definition" message will be displayed in the status bar. A fillet cannot be created when the radius is bigger than the individual length of any of the two picked elements.

The Tool Settings box will look like this →
   i. Operating Procedure:
       1. Choose and/or Input Available Options:
           ▪ "Radius" – input the size of the radius for the fillet
           ▪ "Truncate" – choose the way to trim off the ends of the objects that merge with the ends of the two picked elements
               -   None – A fillet is drawn between the two elements, but NEITHER of the picked elements are trimmed so they meet at the end of each side of the fillet
               -   Both – A fillet is drawn between the two elements, and BOTH of the picked elements are trimmed so they meet at the end of each side of the fillet
               -   First – A fillet is drawn between the two elements, but only the FIRST picked element is trimmed so it meet at one end of the fillet, while the second element is NOT trimmed at its end
       2. Pick the 1st and 2nd element to create a fillet in-between.
       3. Reset (right mouse button).

Example: elements with fillet created per "NONE" Truncate method

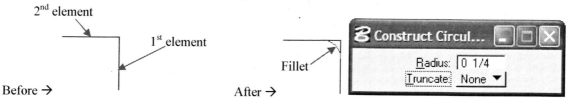

Example: elements with fillet created per "BOTH" Truncate method

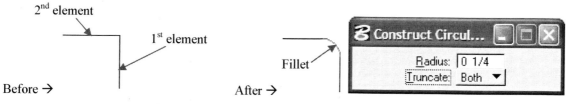

Example: elements with fillet created per "FIRST" Truncate method

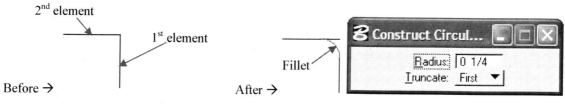

k. *Construct* Chamfer

This tool allows for a chamfer (angled line) to be created between two picked elements. If the size (distance) for the chamfer is bigger than the length of any of the two picked elements, then an "illegal definition" message will be displayed in the status bar. A chamfer cannot be created when the sizes/distances are bigger than the individual length of any of the two picked elements.

The Tool Settings box will look like this →

i. Operating Procedure:
   1. Input Available Options:
       ▪ "Distance 1" – input the size/length to be trimmed off the 1st element, in order to create the chamfer
       ▪ "Distance 2" – input the size/length to be trimmed off the 2nd element, in order to create the chamfer
   2. Enter the chamfer dimensions (Distance 1 and Distance 2) in the Tool Settings box.
   3. Pick the 1st and 2nd element to create a chamfer in-between.
   4. Reset (right mouse button).

Example: elements with ½" x ¼" chamfer

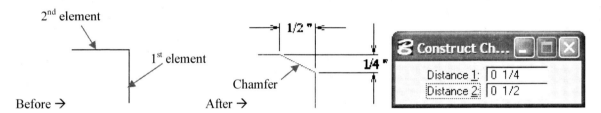

NOTE: *The chamfer dimensions/sizes shown above are for reference only. It will not be shown when the actual chamfer tool is used.*

Chapter 6: Element Attributes

Every man is special, defined by their own individual identity. It is their inherent characteristics (attributes) that separate one person from the next. Every element in a drawing can also be characterized by its attributes, such as Color, Style, Weight, and Level. As with humans that sometime share attributes with each other, the same can be said of elements in a drawing. As we grow older, some of our characteristics are changed by external forces. Similarly, elements attributes are sometimes required to be changed or matched as a drawing develops. After completing this chapter, you'll be able to demonstrate understanding of the following:

- Color
- Style
- Weight
- Level
- Change Attributes
- Match Attributes

I. Setting Element Attributes

Every element placed in a drawing has 4 individual attributes (characteristics) that describes them. They are Color, Style, Weight, and Level. An element can be made of a green Color, with a hidden line Style, that's of a medium wide Weight thickness, and on a specific level.

Of all four attributes, Level is not intuitively understood by the new Microstation user. Levels are like layers in AutoCAD. Consider a level as a clear sheet of paper that has objects drawn on it. A drawing can be made up of a set of many levels. Different groups of elements can be placed on any specific level so they can be turned on or off (temporarily removed from drawing) without affecting other areas of the drawing. For example: *if developing plans/drawings to get a house built, piping and plumbing can be placed on the "PLUMBING" level, electrical components on the ELECTRICAL" level, and walls on the "FLOORPLAN" level. The "PLUMBING" level can be printed with the "FLOOR PLAN" level to see how the pipes will be laid out in the house. If the Levels feature wasn't available, the drawing would be cluttered and hard to read.* The main area to define and adjust level settings is located at **Settings** pulldown menu> **Level** > **Manager**

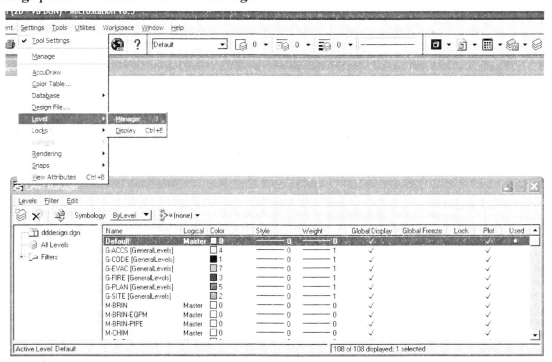

Each Level can have its own specific Color, Style, and Weight setting for elements placed on that level. Level creation and manipulation will be discussed more in *Microstation v8: Simplified*, Vol. 1, Part B. Nonetheless, the following is a table listing the number (quantity) of each attributes available or can be created throughout the use of Microstation:

Attribute	Qty. (Range)	Qty. (Total)	Example
Color	0 to 255	256	4 (Yellow)
Style	0 to infinity	Unlimited*	5 (Hidden)
Weight	0 to 31	32	15 (Thick)
Level	0 to infinity	Unlimited*	Plumbing

* Additional Styles and Levels can be created, so there is no limit on their quantities.

Each of these attributes can be set before the element is placed and changed after it is placed. A default setting for all four attributes is always active (current) and displayed in the Attributes toolbar (below the pulldown menus at the top of the Microstation screen), which is accessed by going to the **Tools** pulldown menu> **Atttibutes**. *NOTE: a checkmark beside Attributes means that the toolbar should be visible.*

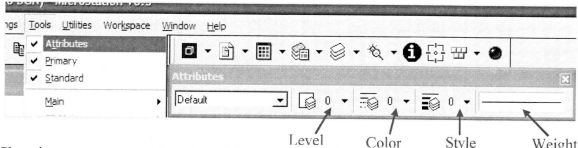

Changing any one or combination of the attributes in the Attributes toolbar will affect those specific characteristics of the "next" element drawn. However, selecting an element (using Element Selection or Fence) and then changing any one or combination of the attributes in the Attributes toolbar will affect those specific characteristics of the "selected" element.

Best practices suggest that a drawing be planned out before starting, and continuing as follows:
- Determine Levels in which to place elements on
- Setup those Levels with their own Color, Style, and Weight attributes in the Level Manager dialog box
- Choose the appropriate Level in the Attributes toolbar
- Choose "By Level" for the Color, Style, and Weight in the Attributes toolbar. *NOTE: "By Level" denotes that the active (current) Color, Style, and Weight attribute will be based off those setup in the Level Manager dialog box*
- Start drawing

II. Changing Element Attributes
After a drawing is started, changes to the attributes of existing elements may be required. The "nominal" way to make the changes is through the use of the **Change Element Attributes** tool in the Change Attributes sub-palette/toolbox, located in the Main Palette.

Change Element
Attributes

The Tool Settings box will look like this →

a. Operating Procedure:
1. Choose Available Options:
    ▪ "Method" – choose the way in which the change/modify an element's attribute
        - Change – changes an element's attribute instantly
        - Match/Change – an existing element's attribute is captured (matched), and is used to change another element's attributes
    ▪ "Use Active Attributes" – place a checkmark beside the feature so the element attributes used for the change process will be based off that shown in the "Attributes" toolbar
    ▪ "Level" – place a checkmark beside the feature and choose the level in which the element should be on
    ▪ "Color" – place a checkmark beside the feature and choose the color in which the element should have
    ▪ "Style" – place a checkmark beside the feature and choose the linestyle in which the element should have
    ▪ "Weight" – place a checkmark beside the feature and choose the line weight in which the element should have
    ▪ "Class" – place a checkmark beside the feature and choose the class in which the element should have
    ▪ "Use Fence" – place a checkmark beside the feature (and choose the fence mode from the associated pulldown menu) to use a fence to change attributes of specific or multiple elements
2. Place a check beside the attribute you want to change.
3. (For "Change" method ONLY) – Pick the element you want to change.
4. (For "Match/Change" method ONLY) – Pick the element you want to "match" (the source).
5. (For "Match/Change" method ONLY) – Pick the element you want to "change" (the destination).
6. ***NOTE:*** *keep picking other element(s) to change, until finished.*
7. Reset (right mouse button).

Example: Green, think line "CHANGED" to a red, thick dashed/hidden line.

Before → ⸻⸻⸻   After → ▀▀ ▀▀ ▀▀ ▀▀ ▀▀ ▀▀ ▀▀

Example: Red, thick dashed/hidden line "MATCHED" and "CHANGED" to a green, think line.

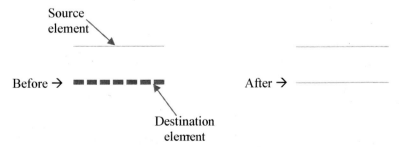

Chapter 7: Text

The tongue can be mightier than the sword. Like the tongue, a drawing can effectively convey information, but through the use of graphics and words called Text. It communicates data not fully explained by dimensions. Text is used in various places in a drawing, such as in the title block, notes, details, and bill of materials. All text are not the same, so understanding how to set various text styles is important, along with how to modify it if incorrectly specified. After completing this chapter, you'll be able to demonstrate the following text associated techniques:

- Text Attributes/Settings
- Placing Text
- Placing Notes
- Modifying Text
- Copy/Increment
- Spell Checker

I. Text

Text is one or more characters combined together to form a word, which will be considered a text string or text node. It is usually understood as a text string when all text fit on one line, but a text node when it captures more than one line. Nonetheless, all text related functions are available in the Text sub-palette/toolbox, which is accessed from the Main Palette.

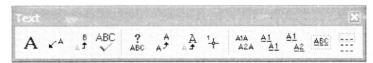

NOTE: *Text Nodes ⊹ and Data Fields (the last four tools in the Text sub-palette/toolbox above) will be covered in <u>Microstation v8: Simplified</u>, Vol. 1, Part B.*

As with an element, Text have attributes and characteristics that need to be setup before they are used, in order to achieve the desired result. Text settings and attributes are accessed by going to the **Element** pulldown menu> **Text Styles.**

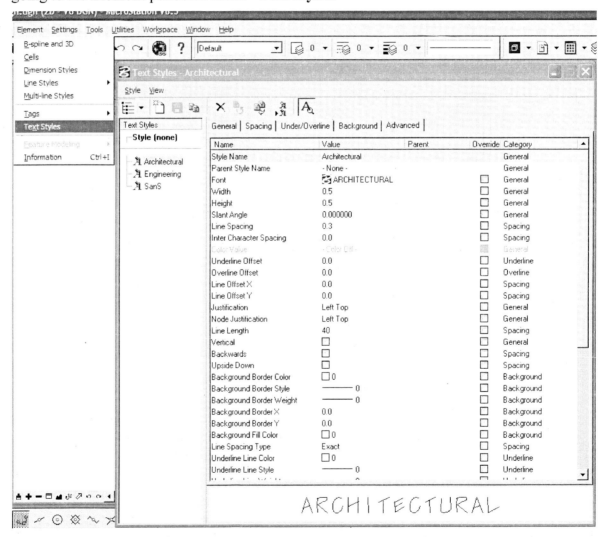

In the Text Styles window, text style settings can be adjusted via the pulldown menus, icons, and tabs. The pulldown menus and icons control the features such as the creation, deletion, and renaming of text styles, while the tabs control the attributes related to the each text style.

Some options for text for each text style are as follows:
1. Font
2. Height and Width
3. Justification
4. Line Spacing and Length
5. Slant (Italicized)
6. Underline
7. Etc.

Once the settings have been set, it is time to start using the text tools.

a. Placing Text **A**

This is the default text tool used for the placing text in the drawing. When this tool is invoked/picked, a Text Editor window will open (along with the Tool Setting tool box).

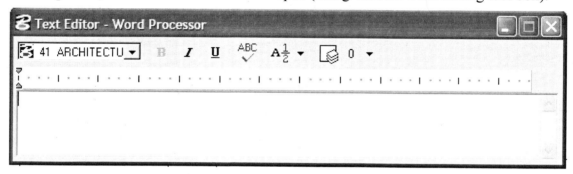

In the Text Editor window, the text to be placed in the drawing will be visible, along with text options such as:
1. Font
2. Bold, Italics, and Underline
3. Spelling (checker)
4. Stacked Fractions
5. Color

NOTE: *Text letter thickness is based on the active line weight.*

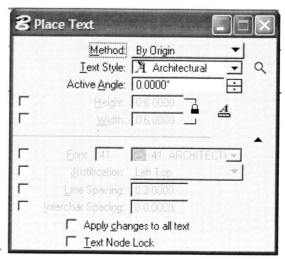

The Tool Settings box will look like this →

i. Operating Procedure:
1. Choose and/or Input Available Options:
- ▪ "Method" – choose the way in which to place text
  - - By Origin – text is placed per the location picked in the design
  - - Fitted – text is fitted inside a specified area, per points placed in the design
  - - View Independent – text placed independent of view orientation
  - - Fitted VI – combination of Fitted and View Independent
  - - Above Element – text is placed above an element
  - - Below Element – text is placed below an element
  - - On Element – text is placed on top of an element
  - - Along Element – text is placed along the path of an element
  - - Word Wrap – long text is fitted inside a specified area (per points placed in the design), by wrapping the lines of the text to fit in the space
- ▪ "Text Style" – choose the text style to use
- ▪ "Active Angle" – input the angle for the text placement
- ▪ "Height" – input the height of the text
- ▪ "Width" – input the width of the text
  *(**NOTE:** 🔒 means the text height and width size will be the same, but different/independent if 🔓 is pictured.)*
- ▪ "Font" – choose the font to use for the text
- ▪ "Justification" – choose the justification for the text
- ▪ "Line Spacing" – input the spacing between each line of the text *(**NOTE:** this option not available with "By Origin" method of text placement)*
- ▪ "Interchar. Spacing" – input the spacing between each character of the text *(**NOTE:** this option not available with "By Origin" method of text placement)*
- ▪ "Apply changes to all text" – place a checkmark beside the feature so all settings inside the Tool Setting Box will affect all existing text in the drawing
- ▪ "Text Node Lock" – place a checkmark beside the feature so every subsequent text will be place in an empty text nodes
2. Enter text in the Text Editor window.
3. Pick a point in the design where the text will be placed
4. *Optional:* Pick other locations in the design where the same text can be placed
5. Reset (right mouse button)

Example: various texts placed per the Above, Below, On, and Along Element methods.

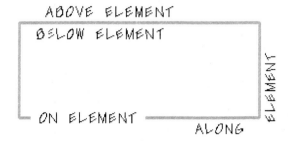

b. Place Notes

This tool allows for text (Notes) to be placed at the end of an arrow with/without a horizontal line called a LEADER. Notes are considered as dimensions because it gives notifications such as: radius, chamfer, slot dimensions, and detail identification.

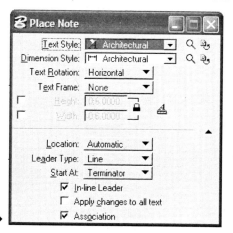

The Tool Settings box will look like this →

i. Operating Procedure:
1. Choose and/or Input Available Options:
   - "Text Style" – choose the text style to use
   - "Dimension Style" – choose the dimension style to use
   - "Text Rotation" – choose the orientation of the text
   - "Text Frame" – choose the frame style of the text
   - "Height" – input the height of the text
   - "Width" – input the width of the text
     (**NOTE:** ⌂ *means the text height and width size will be the same, but different/independent if* ⬛ *is pictured.*)
   - "Location" – choose how the text will be finally positioned
   - "Leader Type" – choose the leader style to be used with the arrow
   - "Start At" – choose where the Note will start from when placed
   - "In-line Leader" – place a checkmark beside the feature so the leader will be placed in line with the text of the Note
   - "Apply changes to all text" – place a checkmark beside the feature so all settings inside the Tool Setting Box will affect all existing text in the drawing
   - "Association" – place a checkmark beside the feature so the Note will be attached to the element, so it reacts with the element when manipulated or modified.
2. Enter text in the Text Editor window.
3. Pick a point in the design (usually ON an element) where the Note will be attached.
4. Pick a point in the design where the text will be placed.
5. Reset (right mouse button).

Example: a Note (with "In-line Leader") is placed on a chamfer.

     c. Display Text Attributes

This tool allows for the attributes associated with picked/existing text to be displayed. *There are NO Tool Settings box options available with this tool.*

      i. Operating Procedure:

        1. Choose or Input Available Options: *NONE*

        2. Pick the text, as the attributes will be displayed in the Status Bar at the bottom of the application window.

Example: the word "TEXT" is picked to display its attributes.

Status Bar

II. Modifying Text

After a drawing is started, adjustments to existing text may be required. They can be checked for correct spelling, edited, changed to new text settings or matched to existing text in the drawing.

     a. Spell Checker

This tool allows for the text in the design to be checked and compared to that of the Microstation dictionary, and suggestions are given if a word is wrong.

The Tool Settings box will look like this →

      i. Operating Procedure:

        1. Choose Available Options:

          ■ "Use Fence" – place a checkmark beside the feature (and choose the fence mode from the associated pulldown menu) to use a fence to spell-check specific or multiple texts

        2. Pick the text to be checked.

        3. Accept (left mouse button).

        4. In the Spellchecker dialog box that opens, make the necessary changes to the incorrect word.

Example: the word "TECT" is being checking for correct spelling.

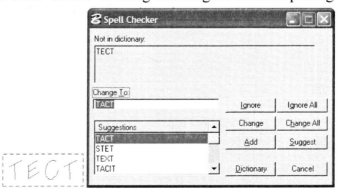

b. Edit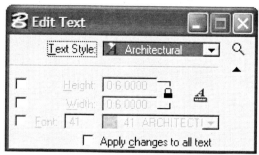
   This tool allows for changing the information in existing text to something else.

The Tool Settings box will look like this →
   i. Operating Procedure:
      1. Choose and/or Input Available Options:
         - "Text Style" – choose the text style to use
         - "Height" – input the height of the text
         - "Width" – input the width of the text
           (**_NOTE:_** 🔒 *means the text height and width size will be the same, but different/independent if ⬚ is pictured.*)
         - "Font" – choose the font to use for the text
         - "Apply changes to all text" – place a checkmark beside the feature so all settings inside the Tool Setting Box will affect all existing text in the drawing
      2. Pick the text to be edited.
      3. In the Text Editor window that opens, change the text information as needed.
      4. Accept (left mouse button).

   Example: the word "TEXT" edited to be "TEST".

   Before → **TEXT**   After → **TEST**

c. Match
   This tool allows for changing the "current/active" text attributes to the attributes of an existing text. It captures and matches attributes of the existing text.
   i. Operating Procedure:
      1. Choose or Input Available Options: *NONE*
      2. Pick the text whose attributes is to be captured/copied ("change to").
      3. Accept (left mouse button).
      NOTE: *the change is not visible until new text is placed, where it will have the attribute(s) of one of the existing texts.*

d. Change
   This tool allows for changing an existing text's attributes to that of the "current/active" text attributes, or any specific attribute.

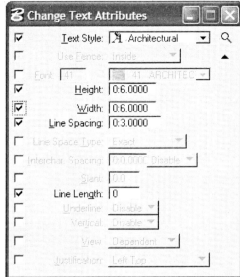

The Tool Settings box will look like this →

i. Operating Procedure:
1. Choose and/or Input Available Options:
    ▪ "Text Style" – choose the text style to use
    ▪ "Use Fence" – place a checkmark beside the feature (and choose the fence mode from the associated pulldown menu) to use a fence to change specific or multiple texts
    ▪ "Font" – choose the font to use for the text
    ▪ "Height" – input the height of the text
    ▪ "Width" – input the width of the text
      (**_NOTE:_** 🔒 _means the text height and width size will be the same, but different/independent if_ 🔓 _is pictured._)
    ▪ "Font" – choose the font to use for the text
    ▪ "Line Spacing" – input the spacing between each line of the text (**_NOTE:_** _this option not available with "By Origin" method of text placement_)
    ▪ "Line Space Type" – choose the type of line spacing for the text
    ▪ "Interchar. Spacing" – input the spacing between each character of the text (**_NOTE:_** _this option not available with "By Origin" method of text placement_)
    ▪ "Slant" – choose to italicize the text
    ▪ "Line Length" – choose the maximum length of the text
    ▪ "Underline" – choose to underline the text
    ▪ "Vertical" – choose to make each character in the text to be vertical
    ▪ "View" – choose to make the text dependent on the orientation of the view
    ▪ "Justification" – choose the justification for the text
2. Pick the text(s) to be changed.
3. Accept (left mouse button).

Example: the word "TEST" changed to have a different font, height, and width.

Before → **TEST**   After → TEST

e. Copy/Increment

The number in a text can be incremented to the next consecutive number, dependent on the value defined for Tag Increment. ***NOTE:*** *You cannot increment letters.*

The Tool Settings box will look like this →

i. Operating Procedure:
   1. Input Available Options:
      ▪ "Tag Increment" – input the increment value
   2. Pick the base text(s) to be copied.
   3. Pick location(s) in the drawing to place the copied and incremented text.
   4. Reset (right mouse button).

Example: the word "TEST1" copied and incremented by 3.

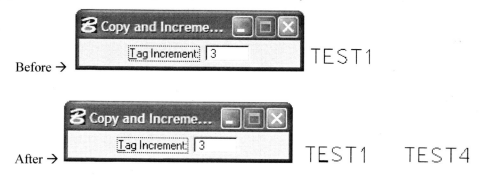

Before →

After →

Chapter 8: Measuring

The character of a man is measured by the accuracy of his truth. Measuring the design is finding out if it is drawn correctly. Every element in the drawing must be geometrically accurate per the desired size. An incorrectly drawn design will result in a part being made too big and/or too small in some or all areas. The design will have to be redrawn and the part remade, resulting in added expense. The preferred solution is to accurately measure the elements during the design process to ensure that it is not made incorrectly. After completing this chapter, you'll be able to demonstrate the following measuring techniques:

- Distance
- Radius
- Angle
- Length
- Area
- Volume

I. Measuring

Measuring elements ensure that the part is drawn accurately. The available measuring tools allow for the determination of distance, radius/diameter, angle, length, area, and volume. The Measure sub-palette/toolbox is accessed from the Main Palette.

a. Measure Distance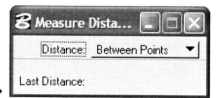

This tool allows for distances to be measured.

The Tool Settings box will look like this →
i. Operating Procedure:
1. Choose Available Options:
    - "Distance" – choose the way in which to measure distance
        - Between points
        - Along an element
        - Perpendicular from an element
        - Minimum between elements
2. For "Between points" Distance Measurement:
    - Pick the start point to be measured.
    - Pick another point to be measured, as a temporary/imaginary line is displayed between the points picked.
    - *The distance is shown in the Tool Settings box and Status bar.*
    - Reset (right mouse button) or continue to pick other points as the length will increase. ***NOTE:*** *Until Reset (right mouse button) is pressed, the distance shown is relative to the 1ˢᵗ point picked.*

Measure Distance Example: measurement between two (2) points in space.

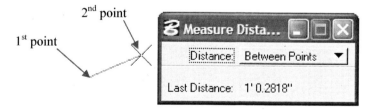

3. For "Along an element" Distance Measurement:
    - Pick the 1ˢᵗ point on an element to be measured.
    - Pick another point on the same element.
    - *The distance is shown in the Tool Settings box and Status bar.*
    - Reset (right mouse button) or continue to pick other points as the length will increase/decrease. ***NOTE:*** *Until Reset (right mouse button) is pressed, the distance shown is relative to the 1ˢᵗ point picked.*

<u>Measure Distance Example:</u> measurement between two (2) points along a smartline.

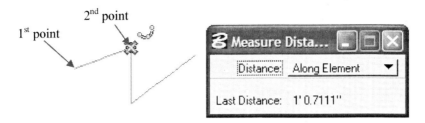

4. For "Perpendicular from an element" Distance Measurement:
   - Pick the 1st element to be measured from.
   - Pick a point on another element, as a temporary/imaginary perpendicular distance line (see arrow) will be displayed from the first element.
   - *The distance is shown in the Tool Settings box and Status bar.*
   - Reset (right mouse button) or continue to pick other points.

<u>Measure Distance Example:</u> perpendicular measurement from one line to another.

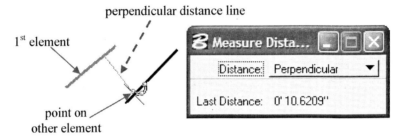

5. For "Minimum between elements" Distance Measurement:
   - Pick the 1st element to be measured from.
   - Pick the other element to measure to, as a temporary/imaginary minimum distance line (see arrow) will be displayed, demonstrating the shortest distance.
   - *The distance is shown in the Tool Settings box and Status bar.*
   - Reset (right mouse button) or continue to pick other points.

<u>Measure Distance Example:</u> minimum measurement between two (2) lines.

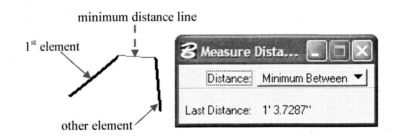

b. Measure Radius
   This tool allows for the radius of an arc, circle, partial ellipse, or full ellipse to be measured.
   There are NO Tool Settings box options available with this tool.

i. Operating Procedure:
   1. Choose and/or Input Available Options: *NONE*
   2. Pick the element you want to measure.
   3. *The radius and diameter is shown in the Tool Settings box and Status bar. For the ellipse, the primary (major) and secondary (minor) radius and diameter will be displayed too.*
   Example: radius measurement of an arc.

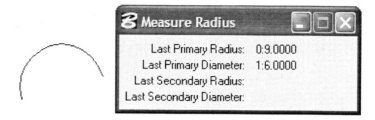

c. Measure Angle
   This tool allows for the angle between two elements to be measured.
   There are NO Tool Settings box options available with this tool.
   i. Operating Procedure:
      1. Choose and/or Input Available Options: *NONE*
      2. Pick the 1st element.
      3. Pick a 2nd element.
      4. Accept (left mouse button).
      5. *The angle is shown in the Tool Settings box and Status bar.*
      Example: angle measurement between two (2) lines.

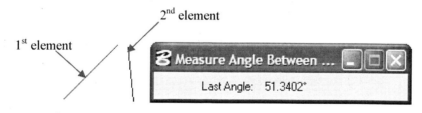

d. Measure Length
   This tool allows for the length of an element (linear or circular) to be measured.

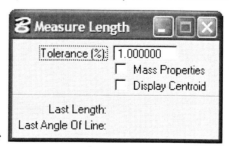

The Tool Settings box will look like this →
   i. Operating Procedure:
      1. Choose and/or Input Available Options:
         ■ "Tolerance %" – input a percentage to use to determine the accuracy of the length for a curved element. Lower is more accurate.

- "Mass Properties" – place a checkmark beside the feature so a second window appears where the unit mass data for a specific material in inputted, resulting in information displayed for the centroid and center of mass of the element
- "Display Centroid" – place a checkmark beside the feature so the centroid (center mark) is displayed in the center of the element

2. Pick the element to be measured
3. Accept (left mouse button).
4. *The length and angle (if element is just a simple line as shown in example below) is shown in the Tool Settings box and Status bar. If Mass Properties is checked, a second window appears with that information, along with Centroid and Center of mass. If Display Centroid is picked, the centroid (center mark) is displayed in the center of the element.*
<u>**NOTE:**</u> *Mass Properties and Display Centroid demonstrated in Measure Area.*

<u>Example</u>: length measurement of a single line.

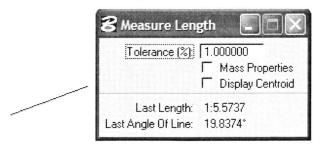

e. Measure Area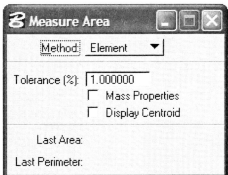
This tool allows for the area of an enclosed element to be measured, dependent on the method being used.

The Tool Settings box will look like this →

i. Operating Procedure:
1. Choose and/or Input Available Options:
- "Method" – choose the way in which to measure the area of an element
    - Element – measurement displayed per the element picked
    - Fence – measurement displayed per the fence in the drawing
    - Intersection – measurement displayed per intersection between two (2) elements picked
    - Union – measurement displayed per joining/combination (union) of two (2) elements picked

- Difference – measurement displayed per subtraction (difference) of one element to another
- Flood – measurement displayed per the inside of an enclosed space
- Points – measurement displayed per points picked, which creates an enclosed space
- "Tolerance %" – input a percentage to use to determine the accuracy of the area for a curved element. Lower is more accurate.
- "Mass Properties" – place a checkmark beside the feature so a second window appears where the unit mass data for a specific material in inputted, resulting in information displayed for the centroid and center of mass of the element
- "Display Centroid" – place a checkmark beside the feature so the centroid (center mark) is displayed in the center of the element
- "Locate Interior Shapes" (Flood method ONLY) – place a checkmark beside the feature so internal elements/shapes can be used in determining area
- "Dynamic Area" (Flood method ONLY) – place a checkmark beside the feature so the mouse is moved to dynamically switch between what enclosed area is measured
- "Max Gap" (Flood method ONLY) – input the maximum allowed space between elements to consider the area as an enclosed shape to be measured

2. For Area Measurement ("Element" Method):
   - Pick the element to be measured.
   - *The area is shown in the Tool Settings box and Status bar.*

   Example: area measurement of an enclosed shape (with Mass Properties and Centroid shown).

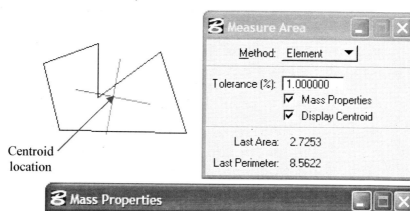

3. For Area Measurement ("Flood" Method):
    ▪ Draw a fence to create the area to be measured.
    ▪ Pick anywhere in the drawing. ***NOTE:*** *The program locates the fence automatically.*
    ▪ *The area is shown in the Tool Settings box and Status bar.*
    Example: area measurement of a fenced area.

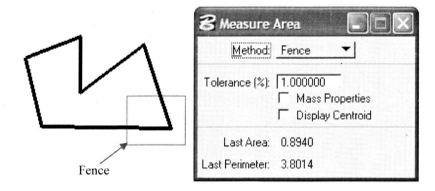

4. For Area Measurement ("Intersection", "Union", or "Difference" Method):
    ▪ Pick the 1st element.
    ▪ Pick the 2nd element.
    ▪ *The area is shown in the Tool Settings box and Status bar.*
    Example: area measurement for intersection between two elements.

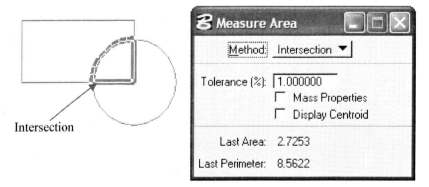

Example: area measurement for union created by two elements.

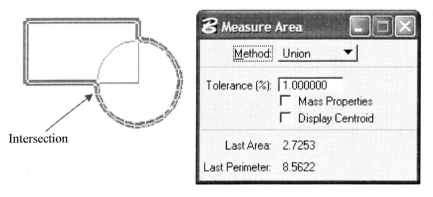

Example: area measurement for difference between two elements.

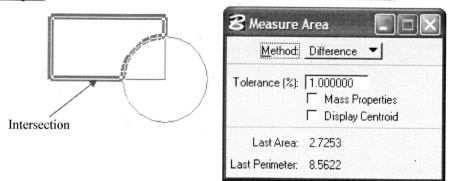

Intersection

5. For Area Measurement ("Flood" Method):
   - Pick inside an enclosed shape.
   - *The area is shown in the Tool Settings box and Status bar.*
   Example: area measurement of enclosed shape (with Locate Interior Shapes).

Interior Shape

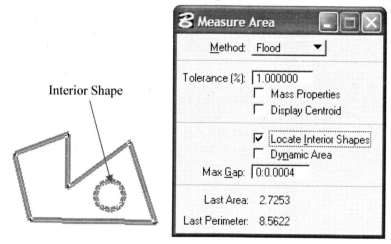

6. For Area Measurement ("Points" Method):
   - Pick points in the drawing creating the area to be measured.
   - Reset (right mouse button).
   - *The area is shown in the Tool Settings box and Status bar.*
   Example: area measurement of enclosed area created from picked points.

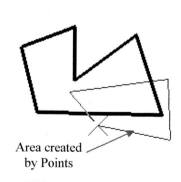

Area created
by Points

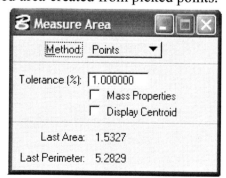

f. Measure Volume
   This is used for 3D drawings only, and won't be covered in this book.

Chapter 9: Dimensioning

Life unguided is one without moral dimensions. A design has little or no use, unless dimensions are shown. Dimensions denote the size in which to make the part. A part cannot be made if the size is not known, and an incorrectly sized part can be costly. Not all elements require the same types of dimensions, as it can vary per style, placement, and/or orientation. Therefore, it is beneficial to understand how to setup and modify dimensions throughout the design process. After completing this chapter, you'll be able to demonstrate the following dimensioning techniques:

- Dimension Attributes/Settings
- Linear Dimensions
- Angular Dimensions
- Radial Dimensions
- Ordinate Dimensions
- Modifying Dimensions

I. Dimensioning

Dimensioning is an essential part of the design process, as it provides information to the manufacturer or builder, so that the part can be made to the accurate size. Microstation provides a variety of dimensioning tools to match all needs. There are two sets of tools available. The nominal Dimensioning sub-palette/toolbox is accessible from the Main Palette.

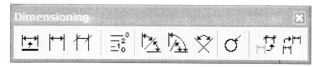

The tools can be subdivided into four sets (Linear, Angular, Radial, and Miscellaneous tools), each corresponding to their functionality. These sets have more advanced tools, and are available in the full Dimension toolbox, accessed by going to the **Tools** pulldown menu> **Dimension Tools**.

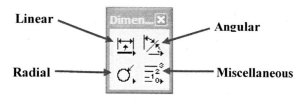

NOTE: *This toolbox includes some of those in the nominal Dimension toolbox, but the others will not be covered in this book.*

As with an element, Dimensions have attributes and characteristics that need to be setup before they are used, to ensure the desired style is achieved. Dimension settings and attributes are accessed by going to the **Element** pulldown menu> **Dimension Styles**.

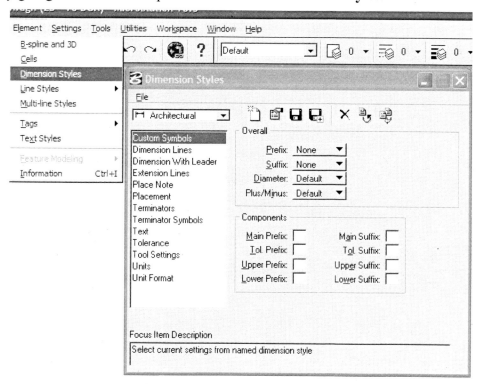

Changes to the Dimension Settings dialog box affects the following dimension attributes:
- Dimension Lines
- Extension Lines
- Text
- Terminator
- Units

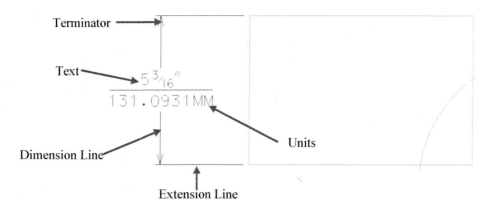

The following describes each feature from the Dimension Styles dialog box.

1. Custom Symbols
- Sets parameters for Custom Symbols for dimensions.

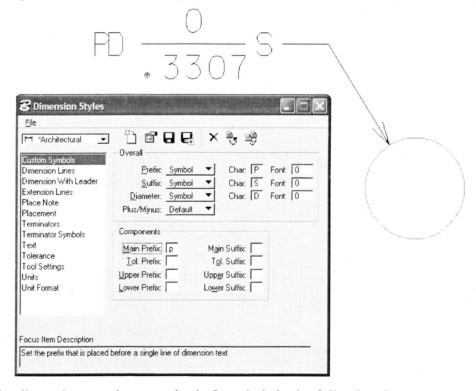

- The dimensions can be comprised of symbols in the following forms:
    - Default – Microstation defined
    - Symbol – a text character
    - Cell – any custom design

2. Dimension Lines
   - Sets parameters for Dimension Lines.

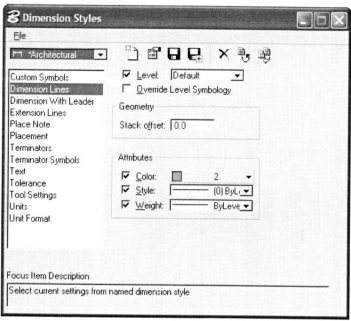

   - The dimension line attributes (Color, Level, Weight, and Linestyle) can be set.
   - The offset distance (between two parallel dimensions based off same origin) of stacked dimension lines can be set.

     NOTE: *The unit used for the number in the "Stack Offset" area is based off the Format specified in the Coordinate Readout section of the Design File dialog box (**Settings** pulldown menu).*

   - Etc.

3. Dimension With Leader
   - Sets parameters for Dimensions with Leader.

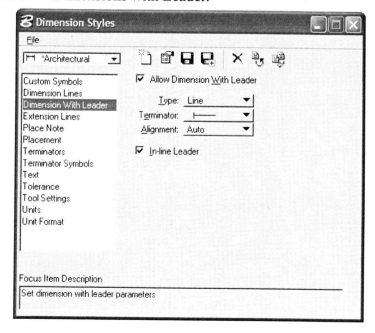

- If "Allow Dimension With Leader" is checked, a leader in the middle of the dimension line will be shown when the dimension is placed. For example:

- The leader line type, terminator style, and alignment can be set.
- Etc.

4. Extension Lines
   - Sets parameters for Extension Lines.

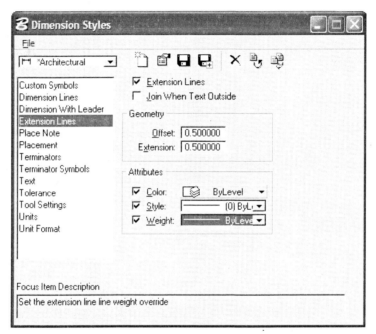

- The extension line attributes (Color, Weight, and Linestyle) can be set:
- Two geometry parameters of the extension line can be set:
  - "Offset" – the distance between object being dimensioned and the start of the extension line.
  - "Extension" – the distance from the dimension line to the end of the extension line.
- If "Extension Lines" is checked, the extension line will be shown when the dimension is placed.
- Etc.

5. Place Note
   ▪ Sets parameters for the placement of Notes.

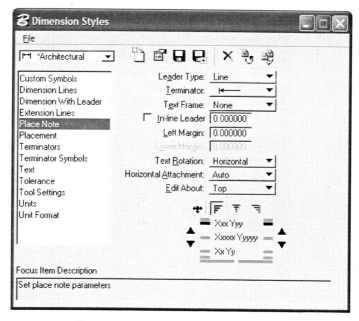

   ▪ The leader line type, terminator style, and framing of the Notes can be set.
   ▪ The orientation and location of the Notes, with respect to the leader, can be set.
   ▪ Etc.

6. Placement
   ▪ Sets parameters for the Placement techniques of dimensions.

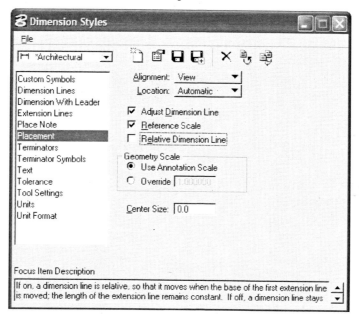

   ▪ The alignment/orientation (View axis, Drawing axis, Arbitrarily, or True) and the
     method used to locate/position (Manual, Semi-automatic, or Automatic by the program)
     of a typical dimension can be set.
     NOTE: *True alignment is recommended for most cases.*

- The size of center marks for circular elements can be set.
  NOTE: The unit used for the number in the "Center Mark" area is based off the Format specified in the Coordinate Readout section of the Design File dialog box (Settings pulldown menu).
- Etc.

7. Terminators
   - Sets parameters for Terminators in dimensions.

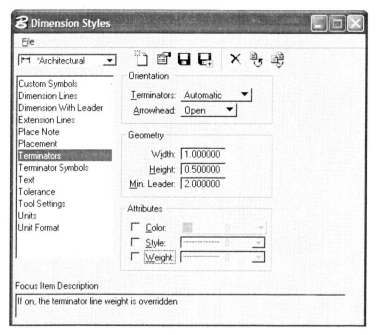

- The style of terminators (Open, Closed, or Filled), and their orientation (inside, outside, reversed, or automatic by the computer), can be set.
- The terminator attributes, such as its color, weight, and linestyle, can be set.
- The size (Width, Height, and Minimum Leader) of the terminators can be set.
  NOTE: The size shown is depicted as a fraction of the text size. For example: a Height of 0.50000 means that the terminator height is half the text height.
- Etc.

8. Terminator Symbol
   - Sets custom Symbols for the various types of Terminators used with the various dimension tools:
     - Arrow
     - Stroke
     - Origin
     - Dot
     - Note

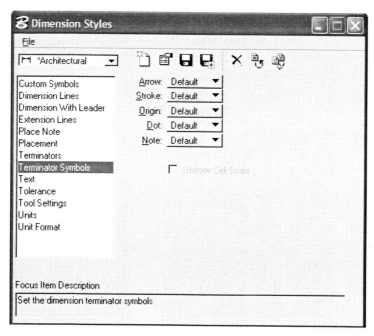

- The various terminators can be represented as symbols in the following forms:
  - Default – Microstation defined
  - Symbol – a text character
  - Cell – any custom design
- Etc.

9. Text
  - Sets parameters for Text used in dimensions.

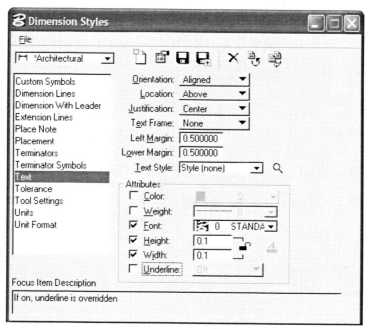

- The Text attributes of a dimension can be based off an existing Text Style, or an otherwise explicit/external Color, Weight, Font, Height, Width, and/or Underline.

- The orientation (Aligned or Horizontal), location (Above or Inline), and justification (Left, Center, or Right) of the dimension Text can be set.
- Etc.

10. Tolerance
   - Sets parameters for Tolerances noted in dimensions.

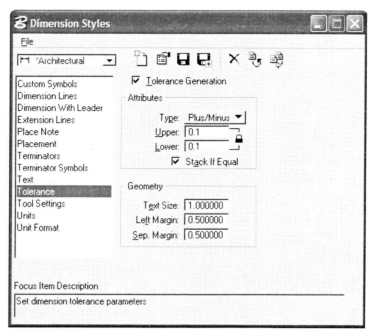

- If "Tolerance Generation" is checked, manufacturing tolerances can be specified and will be shown. For example: If "Type" is Plus/Minus, and Upper and Lower is set to 0.005, a part that is 3.125mm can be manufactured to 3.120 mm to 3.130 mm.
   NOTE: *The unit used for the number in the "Upper" and "Lower" area is based off the Format specified in the Coordinate Readout section of the Design File dialog box (**Settings** pulldown menu).*
- Three geometry parameters of the tolerance notation can be set:
   - Text Size – the size of the text.
   - Left Margin – the margin between the dimension text and the tolerance text.
   - Sep. Margin – the margin between the dimension line and the tolerance text.
   NOTE: *The size of each of the tolerance geometry parameters is depicted as a fraction of the text size. For example: a Text Size of 1.00000 means that the tolerance height is the same (1 to 1) as the text height.*
- Etc.

11. Tool Settings
   ▪ Sets which parameters are visible in each dimension Tool.

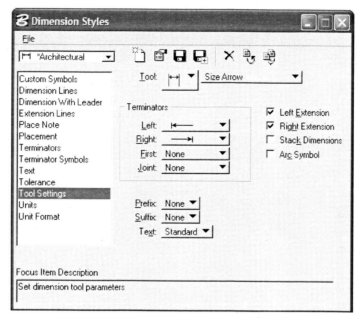

   ▪ How each dimension tool works can be customized. *For example, the Left Extension line for the "Size Arrow" dimension tool can be turned off, or the Left terminator can be chosen to be a stroke and the Right terminator an arrow.*
   ▪ Etc.

12. Units
   ▪ Sets parameters for the Units shown in dimensions.

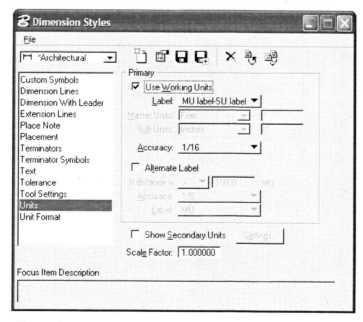

   ▪ The Units of a dimension can be based off the Working Units (**Settings** pulldown menu> **Design File**> **Working Units**), or an otherwise explicit/external Master and Sub Unit format.

- The accuracy of the dimension can be set. *For example: mm can be limited to 0.01 instead of 0.0001, so 1.2317 would be 1.23.*
- If "Show Secondary Units" is checked, a different unit can be specified to be shown below the Primary Unit when the dimension is placed. The Secondary Unit is different, yet equivalent o the Primary Unit. For example: While the Primary units are set to inches, Secondary units could be set to millimeters.
- "Alternate Label" is used when a different label is needed to be shown when a certain criteria have been met.
- The "Scale Factor" of the dimensions ONLY can be set.
  NOTE: *This feature is effective because the part can be drawn at a certain scale, and the dimensions can be set to a different scale, relinquishing the need to redraw or manipulate (manually scale) the part.*
- Etc.

13. Unit Format
    - Sets parameters for dimension unit format.

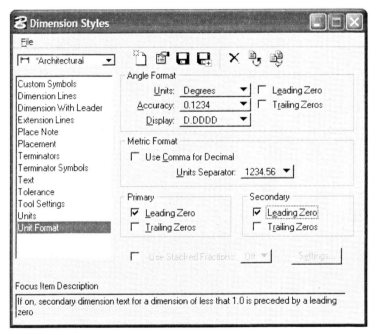

- The format of the degrees can be set:
  - Unit type
  - Accuracy
  - Display style
- The features of Metric dimensions can be set.
- The visibility of the Zeroes in the front (Leading) and back (Trailing) of the dimensions can be set. *For example: A Leading zero is 0.005", and a trailing zero is 1.250".*
- Etc.

Once the settings have been set, it is time to start using the dimension tools.

a. Element Dimensioning
   This tool is a smart or intelligent tool that identifies what shape/element is being
   dimensioned, then it determines and creates the appropriate dimension for it (as the Tool
   Setting box display changes accordingly).

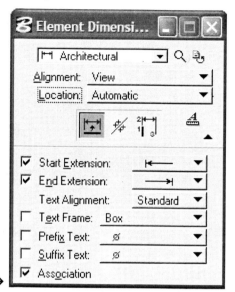

The Tool Settings box will look like this →

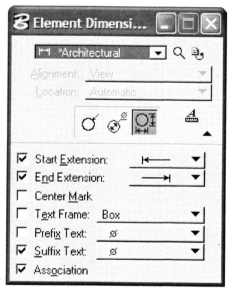

Or look like this →
   i. Operating Procedure:
      1. Choose Available Options:
         ▪ Element Dimension Tools:
            - Dimension Element
            - Label Line
            - Dimension Size Perp - Pts
            - Dimension Radius
            - Dimension Diameter (Extended)
            - Dimension Diameter Parallel
         ▪ "Dimension Style" – choose the preset style to use for the dimension

- ▪ "Alignment" – choose the way to align the dimension when placed
  - View
  - Drawing
  - True
  - Arbitrary
- ▪ "Location" – choose the way to finish positioning the dimension when placed
  - Automatic
  - Semi-Auto
  - Manual
- ▪ "Start Extension" – place a checkmark beside the feature and choose which terminator style to use for the start extension line
- ▪ "End Extension" – place a checkmark beside the feature and choose which terminator style to use for the end extension line
- ▪ "Center Mark" – place a checkmark beside the feature and a check mark will be shown in the dimension
- ▪ "Label Line" – choose a label format and location on a line
  - Length/Angle
  - Angle /Length
  - Length Above
  - Angle Above
  - Length Below
  - Angle Below
  - Length Angle Above
  - Length Angle Below
- ▪ "Text Alignment" – choose the alignment strategy for the text in dimensions
- ▪ "Text Frame" – place a checkmark beside the feature and choose the frame style for the Text ·
- ▪ "Prefix Text" – place a checkmark beside the feature and choose the prefix notation for the text in dimension
- ▪ "Suffix Text" – place a checkmark beside the feature and choose the suffix notation for the text in dimension
- ▪ "Association" – place a checkmark beside the feature so the Dimension will be attached to the element, so it reacts with the element when manipulated or modified.

2. For "Dimension Element" ⊞
   - ▪ Pick the element, and the dimension will be drawn.
   - ▪ Pick a location to place the dimension.

Example:

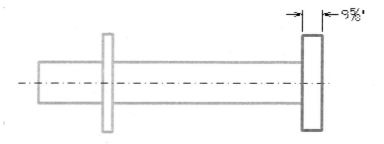

3. For "Label Line"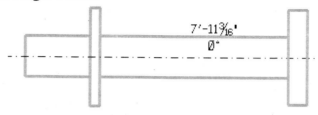
   - Pick a point on an element, and the dimension label will be drawn.
   - Pick a location to place the dimension.

   Example: Length/Angle format

4. For "Dimension Size Perp - Pts"
   - Pick a point on an element to be perpendicular from, and the dimension will be drawn.
   - Pick another point.

   Example:

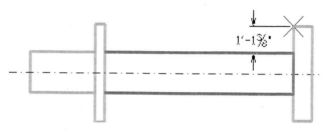

5. For "Dimension Radius"
   - Pick the element, and the dimension will be drawn.
   - Pick a location to place the dimension.

   Example:

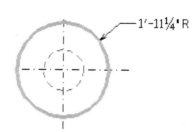

6. For "Dimension Diameter (Extended)"
   - Pick the element, and the dimension will be drawn.
   - Pick a location to place the dimension.

   Example:

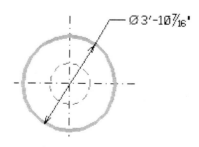

7. For "Dimension Diameter Parallel"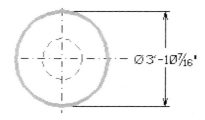
    - Pick the element, and the dimension will be drawn.
    - Pick a location to place the dimension.

    Example:

Ø 3'-10⁷/₁₆'

b. Linear Dimensioning
    These tools create dimensions that depict lengths.

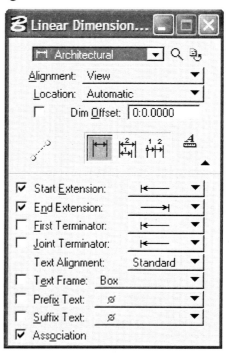

The Tool Settings box will look like this →

i. Operating Procedure:
    1. Choose and/or Input Available Options:
        - Linear Dimension Tools:
            - Dimension Linear Size
            - Dimension Location Stacked
            - Dimension Location
        - "Dimension Style" – choose the preset style to use for the dimension
        - "Alignment" – choose the way to align the dimension when placed
            - View
            - Drawing
            - True
            - Arbitrary

- "Location" – choose the way to finish positioning the dimension when placed
  - Automatic
  - Semi-Auto
  - Manual
- "Dim Offset" – input a size for the spacing between each dimension
- "Start Extension" – place a checkmark beside the feature and choose which terminator style to use for the start extension line
- "End Extension" – place a checkmark beside the feature and choose which terminator style to use for the end extension line
- "First Terminator" – place a checkmark beside the feature and choose which terminator style to use for the first terminator
- "Joint Terminator" – place a checkmark beside the feature and choose which terminator style to use for the dimension joints

  NOTE: _For Dimension Linear Size and Dimension Location ONLY_
- "Text Alignment" – choose the alignment strategy for the text in dimensions
- "Text Frame" – place a checkmark beside the feature and choose the frame style for the Text
- "Prefix Text" – place a checkmark beside the feature and choose the prefix notation for the text in dimension
- "Suffix Text" – place a checkmark beside the feature and choose the suffix notation for the text in dimension
- "Association" – place a checkmark beside the feature so the Dimension will be attached to the element, so it reacts with the element when manipulated or modified.

2. For "Dimension Linear Size"
   - Pick the start point of the 1st extension line.
   - Pick the start point of the 2nd extension line.
   - Pick the location of the dimension line, and the dimension will be drawn.
   - Reset (right mouse button), or pick the start point for an adjacent dimension's extension line.

Example:

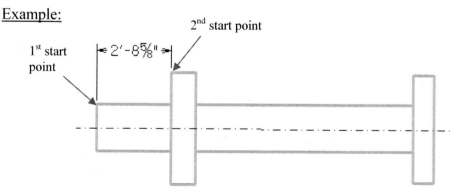

3. For "Dimension Location (Stacked)"
   - Pick the start point of the 1st extension line.
   - Pick the start point of the 2nd extension line.
   - Pick the location of the dimension line, and the dimension will be drawn.

- Pick the start point for a "Stacked" dimension's extension line, and the "Stacked" dimension (overall dimension) will be drawn above the first dimension.
- Reset (right mouse button), or pick the start point for another "Stacked" dimension's extension line.

NOTE: *the distance between the first and second dimension line is set per the Dimensioning topic above (subtopic #2).*

Example:

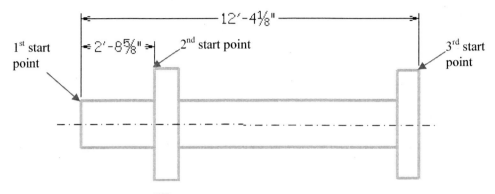

4. For "Dimension Location"
   - Pick the start point of the 1st extension line.
   - Pick the start point of the 2nd extension line.
   - Pick the location of the dimension line, and the dimension will be drawn.
   - Pick the start point for an adjacent dimension's extension line, and the overall dimension will be drawn alongside the first dimension.
   - Reset (right mouse button), or pick the start point for another adjacent dimension's extension line.

Example:

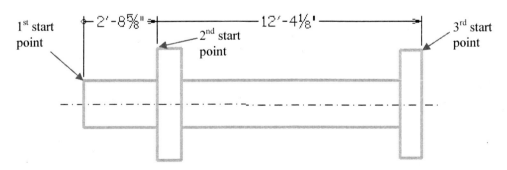

c. Angular Dimensioning
These tools create dimensions that depict angles.

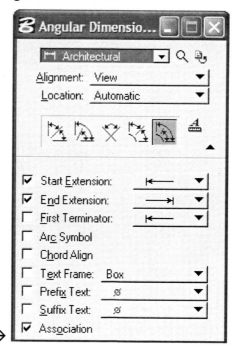

The Tool Settings box will look like this →
i. Operating Procedure:
   1. Choose Available Options:
       ▪ Angular Dimension Tools:
          - Dimension Angle Size
          - Dimension Angle Location
          - Dimension Angle Between Lines
          - Dimension Arc Size
          - Dimension Arc Location
       ▪ "Dimension Style" – choose the preset style to use for the dimension
       ▪ "Alignment" – choose the way to align the dimension when placed
          - View
          - Drawing
          - True
          - Arbitrary
       ▪ "Location" – choose the way to finish positioning the dimension when placed
          - Automatic
          - Semi-Auto
          - Manual
       ▪ "Start Extension" – place a checkmark beside the feature and choose which
                 terminator style to use for the start extension line
       ▪ "End Extension" – place a checkmark beside the feature and choose which
                 terminator style to use for the end extension line
       ▪ "First Terminator" – place a checkmark beside the feature and choose which
                 terminator style to use for the first terminator
       ▪ "Arc Symbol" – place a checkmark beside the feature and an arc symbol will be
                 shown above the dimension

- "Chord Align" – place a checkmark beside the feature and the dimension will be aligned with the chord
- "Text Frame" – place a checkmark beside the feature and choose the frame style for the Text
- "Prefix Text" – place a checkmark beside the feature and choose the prefix notation for the text in dimension
- "Suffix Text" – place a checkmark beside the feature and choose the suffix notation for the text in dimension
- "Association" – place a checkmark beside the feature so the Dimension will be attached to the element, so it reacts with the element when manipulated or modified.

2. For "Dimension Angle Size"
   - Pick the starting point for the 1st extension line.
   - Pick the axis of the angle (where the two lines intersect.
   - Pick the starting point for the 2nd extension line.
   - Pick a location to place the dimension.
   - Reset (right mouse button) to end the dimension command, or pick another extension line starting point to place an adjacent (beside the current/last completed) dimension.

   Example:

   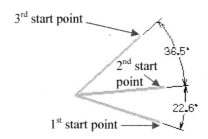

3. For "Dimension Angle Location"
   - Works similar to "Dimension Angle Size", except the dimensions are stacked above each other, not beside each other.

   Example:

   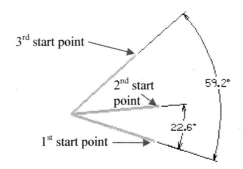

4. For "Dimension Angle Between Lines"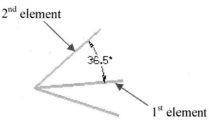
   - Pick the first element.
   - Pick the second element, and the angle between the two elements will be displayed.
   - Pick a location to place the dimension.

Example:

5. For "Dimension Arc Size"
   - Pick the starting point for the 1st extension line.
   - Pick the starting point for the 2nd extension line.
   - Pick a location to place the dimension.
   - Reset (right mouse button) to end the dimension command, or pick another extension line point to place an adjacent (beside the current/last completed) dimension.

Example:

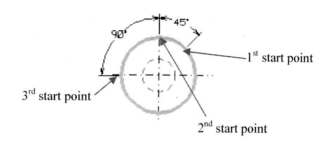

6. For "Dimension Arc Location"
   7. Works similar to "Dimension Arc Size", except the dimensions are stacked above each other, not beside each other.

Example:

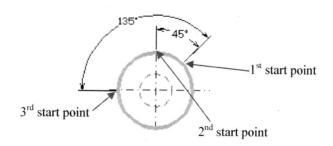

d. Radial Dimensioning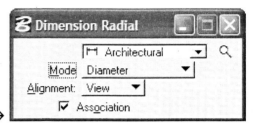
   This tool creates dimensions that depict radius, diameter, and center mark of a circular element.

The Tool Settings box will look like this →

i. Operating Procedure:
   1. Choose Available Options:
      - "Dimension Style" – choose the preset style to use for the dimension
      - "Mode" – choose the type of radial dimension to create
         - Radius
         - Radius Extended
         - Diameter
         - Diameter Extended
         - Center Mark
      - "Alignment" – choose the way to align the dimension when placed
         - View
         - Drawing
         - True
         - Arbitrary
      - "Association" – place a checkmark beside the feature so the Dimension will be attached to the element, so it reacts with the element when manipulated or modified.
   2. Pick the circular element to be dimensioned.
   3. Pick a location to place the dimension.
      "Radius" Mode Example: a circle with a 10-3/16 inches radius.

      —10³⁄₁₆'R

      "Radius Extended" Mode Example: a circle with a 10-3/16 inches radius.

      —10³⁄₁₆'R

      "Diameter" Mode Example: a circle with a 1ft 8-3/8 inches diameter.

      —1'-8³⁄₈'∅
      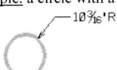

"Diameter Extended" Mode Example: a circle with a 1ft 8-3/8 inches diameter.

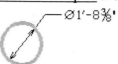

"Center Mark" Mode Example: a circle with a center mark shown

e. Ordinates Dimensioning
   This tool creates dimensions that label distances along an axis from an origin.

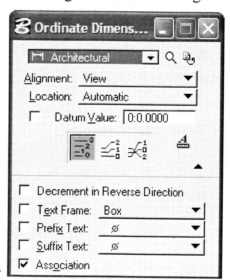

The Tool Settings box will look like this →
   i. Operating Procedure:
      1. Choose Available Options:
         ▪ Ordinates Dimension Tools:
            - Ordinate Unstacked
            - Ordinate Stacked
            - Ordinate Free Location
         ▪ "Dimension Style" – choose the preset style to use for the dimension
         ▪ "Alignment" – choose the way to align the dimension when placed
            - View
            - Drawing
            - True
            - Arbitrary
         ▪ "Location" – choose the way to finish positioning the dimension when placed
            - Automatic
            - Semi-Auto
            - Manual
         ▪ "Datum Value" – the start value of the dimension
         ▪ "Decrement in Reverse Direction" – list the dimension in the opposite direction, which is to decrement

- "Text Frame" – place a checkmark beside the feature and choose the frame style for the Text
- "Prefix Text" – place a checkmark beside the feature and choose the prefix notation for the text in dimension
- "Suffix Text" – place a checkmark beside the feature and choose the suffix notation for the text in dimension
- "Association" – place a checkmark beside the feature so the Dimension will be attached to the element, so it reacts with the element when manipulated or modified.

2. For "Ordinate Unstacked"
   - Pick where the extension line to start from = Ordinate Origin.
   - Pick the direction where all dimensions will be placed or should go.
   - Pick a location to place the dimension, and the dimension will be drawn.
   - Pick the location of the next dimension line, or Reject it.

   NOTE: *All dimensions, while in the command, will be based off the Origin.*

   Example:

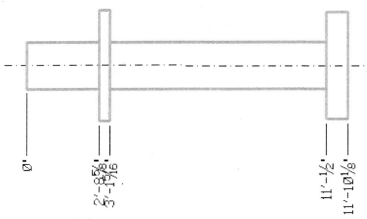

3. For "Ordinate Stacked"
   - Works similar to "Ordinate Unstacked", except the dimensions are stacked above each other if they are too close to each other.

   Example:

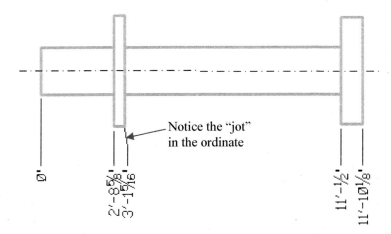

Notice the "jot" in the ordinate

4. For "Ordinate Free Location"
   - Works similar to "Ordinate Unstacked" and "Ordinate Unstacked", except they can be freely positioned.
   NOTE: *The "Location" option should be set to "Manual", and the "Stacked Dimension" setting should be checked in the Dimension Styles dialog box.*

II. Modifying Dimensions

After a drawing is started, adjustments to existing dimensions may be required. They can be changed to new dimension settings or matched to existing dimensions.

a. Change Dimension

This tool allows for changing an existing dimension's attributes to that of the "current/active" dimension attributes.

The Tool Settings box will look like this →
   i. Operating Procedure:
      1. Choose Available Options:
         - "Dimension Style" – choose the preset style to use for the dimension
      2. Pick the dimension whose setting/attributes you want **"changed"**.
      3. Accept (left mouse button).
         NOTE: CHANGE *= matches drawing to attributes =* ***changes drawing.***

b. Match Dimension

This tool allows for changing the "current/active" dimension attributes to the attributes of an existing dimension. It captures and matches attributes of the existing dimension. *There are NO Tool Settings box options available with this tool.*

   i. Operating Procedure:
      1. Choose and/or Input Available Options: *NONE*
      2. Pick the dimension whose setting/attributes you want to **"change to"** or **"match with".**
      3. Accept (left mouse button).
         NOTE: MATCH *= matches attributes to drawing =* ***changes attributes.*** *The change is not visible until a new dimension is placed, of which it will have the attribute(s) of an existing dimensions (that to be "matched with").*

Chapter 10: Printing

Education is the root of all print. The print provides all the necessary information to educate the manufacturer how to make the part or design. The lack of information results in assumptions and speculations. Sufficient details should be incorporated into the print to ensure efficient production. All important parameters should be chosen when setting up the printing process. After completing this chapter, you'll be able to do the following:

- Define the printed area
- Choose the proper paper and printer
- Set up the scale of the print
- Preview the print
- Create a print

I. Printing

The design is the creation of one's idea, which will ultimately be made into something tangible. In order to reach that goal, the design must be delivered to the manufacturer. This is accomplished via an electronic (file) or physical (printed hard copy) distribution. They are one in the same, as the print is the physical representation of the electronic design file. The electronic file depicts the design on the computer monitor, or alternately on printed paper. This is not always a one to one translation. Proper printing techniques ensure that the appropriate information is transferred from the electronic file to the paper hard copy. Printing can be accomplished by easily going to:

- The Printer icon 🖨 (see arrow below) in the Standard tool bar

NOTE: *if the Standard tool bar is not shown, go to the **Tools** pulldown menu> **Standard***

- The **File** pulldown menu> **Print**

- Pressing **Ctrl** and **P** on the keyboard

As a result, the following printing window will open:

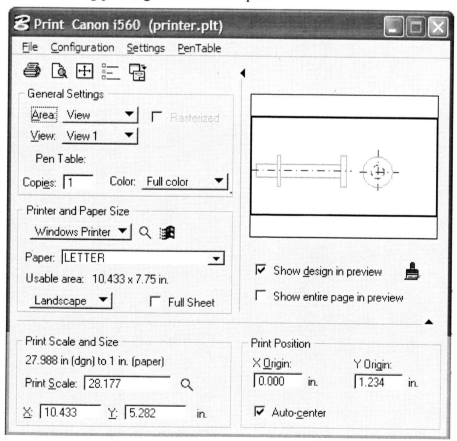

This window allows for setting and changing of printing parameters. Changes in this window will determine how the print is made and what is embodied in the print. The settings should be confirmed before actual printing commence. Settings include, but are not limited to:

- Position of design on the paper

- Print Scale/Size (design to paper ratio)

- Printer and Printer driver

- Paper Size and Orientation

- Pen Table Configuration for Plotter

II. General Settings

These settings are general in function and include parameters such as the number of copies, the color of the print (Monochrome, Grayscale, or Colored), the pen table definitions (if plotter is used), and the printed area.

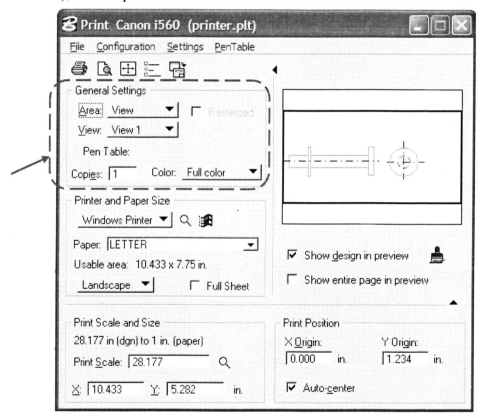

The "Area" setting determines the space in the drawing that will be printed, which can be based upon the:

- *View*'s visible area

- *Fence* in the drawing

- *Sheet*

- *Fit*ted maximum allowed space (extents of the drawing).

The "View" setting allows for any open view to be used as the basis for printing (as long as the "Area" pulldown menu is set to View).

NOTE: *A Pen Table can be created, attached, edited, detached, or imported by going to the **PenTable** pulldown menu in the Print window.*

III. Printer and Paper Size

Once the printed area is determined, the medium (printer) used for printing and the media (paper) to be printed on is chosen. Microstation provides two types of printing medium formats:

- Windows Printer

- Bentley Printer Driver

The Windows Printer will use the standard drivers and settings for a Microsoft Windows based printer. If an alternate output is needed, then use the Bentley derived drivers. The Windows Printer option will typically print directly to a machine, while the Bentley Printer Driver will typically print to a file (i.e.: pdf.plt = Acrobat pdf file). The appropriate Windows Printer or Bentley Printer Driver can be chosen from the icons, ⌕ and 🖳, respectively.

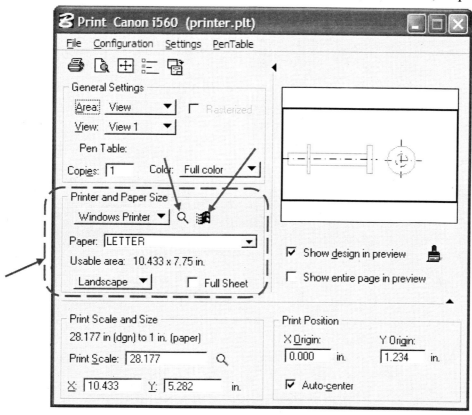

The "Paper" setting determines the paper size to be printed on, as the usable (printable) area on the paper is displayed. The printed orientation (Landscape or Portrait) of the design on the paper is chosen from its pulldown menu. "Full Sheet" allows for use of whole paper to print.

NOTE: *Modification to the Page Setup and/or drivers can be done by going to the **File** pulldown menu.*

IV. Print Scale, Size, and Position

The "Print Scale" is the ratio between the size of the design and the size of the paper. The sizes are described in linear measurement units. The units for the design are based upon the Working Units (Master and Sub Unit) setup for the drawing (the **Settings** pulldown menu> **Design File**> **Working Units** in the main Microstation drawing area/design plane). The unit for the paper is per the **Settings** pulldown menu> **Units** in the Print window.

The Print Size is the maximum size/space occupied by the design on the paper, which will change based upon the print scale chosen. The print size is displayed in terms of the X and Y axis, as it fits on the paper.

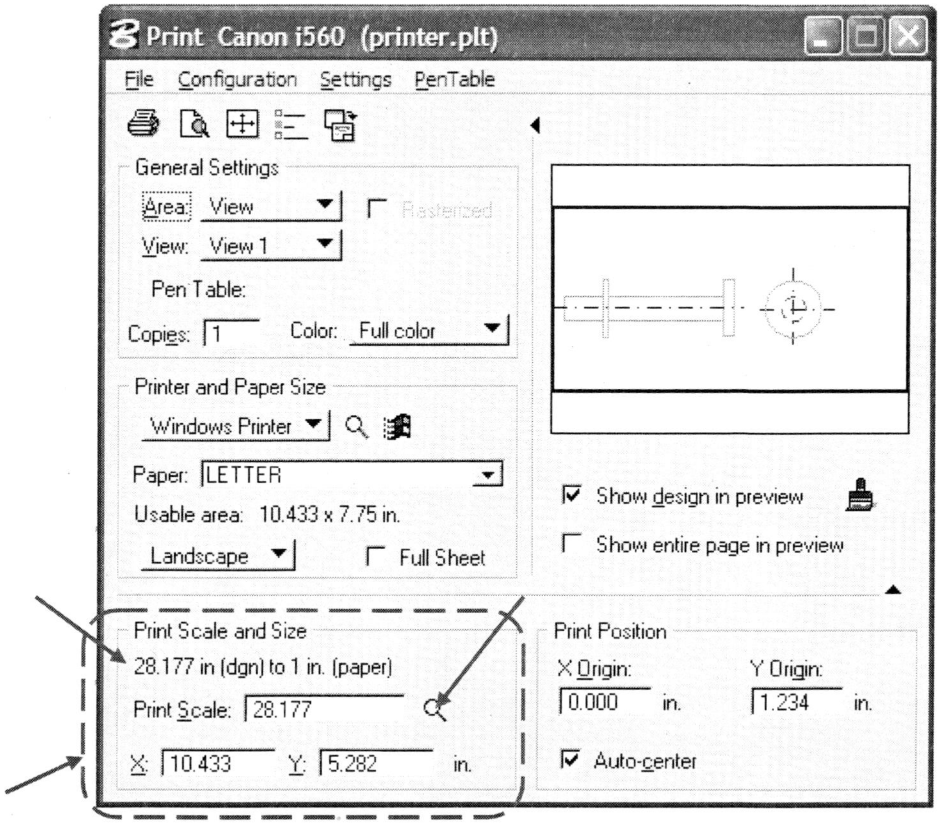

It should be noted that changing the "Area" in the "General Settings" section of the Print window, will result in an initial print scale and size. That initial print scale and size is a default and can be changed thereafter, by picking on the Scale Assistant icon (see arrow above) to open the following dialog box:

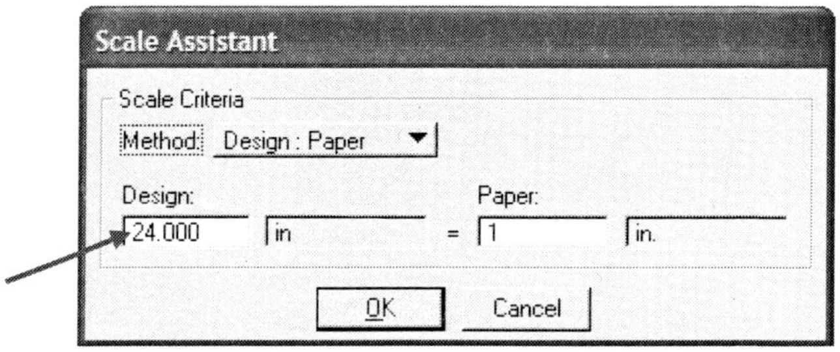

NOTE: *The Scale Assistant (shown above) is suggesting that 1" measured on the paper is equal to 24" of the actual manufactured design.*

As a result of the Scale Assistant change, the drawing to paper scale is changed:

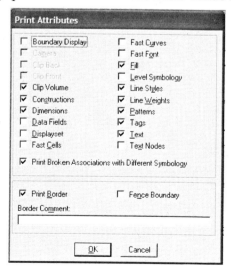

The "Print Position" is the location of the design on the paper (the lower left limit of the design coincident with the origin (0,0), lower left corner, of the paper). Changing the Print Scale and Size will affect the Print Position. "Auto-center" centers the design on the paper.

V. Print Attributes

Most parameters can be adjusted from the various sections of the Print window, and also from pulldown menus. Other parameters can be changed from the pulldown menus, like "Print Atttibutes" from the **Settings** pulldown menu of the Print window:

Placing a checkmark beside any of the attributes (i.e. Patterns) will include it into the print.

VI. Print Preview and Print
The print should be previewed before the final output is initiated to the printer device or file. The Print window allows for the preview to be focused on the design only or the entire printed paper (including the design), by placing a checkmark besides either of the two options:

- Show design in preview

- Show entire page in preview

The preview will be shown in the Print window (see above), or in its own windows by going to the **File** pulldown menu> **Print Preview** (or using the icon in the Print window):

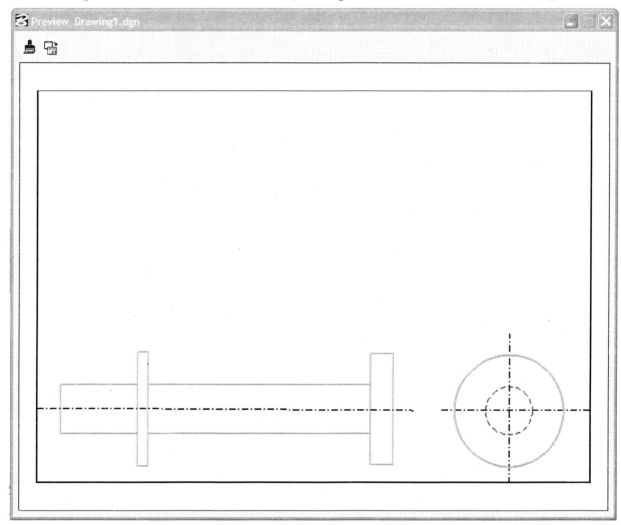

NOTE: *The Print Preview window can be opened directly from with the main Microstation drawing area (design plane), by going to the **File** pulldown menu> **Print Preview**.*

Once the preview as been deemed acceptable, the print can be initiated by using the icon in the Print window, or by going to the **File** pulldown menu> **Print**.

Congratulations! The manufacturing process can now begin.

Appendix A: Projects

A.1: Mechanical

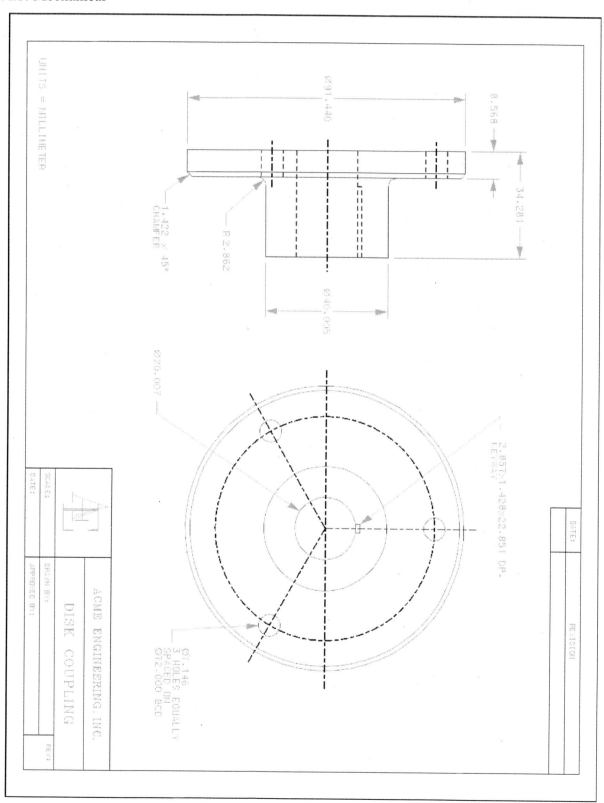

Microstation v8: Simplified

A.2: Electrical

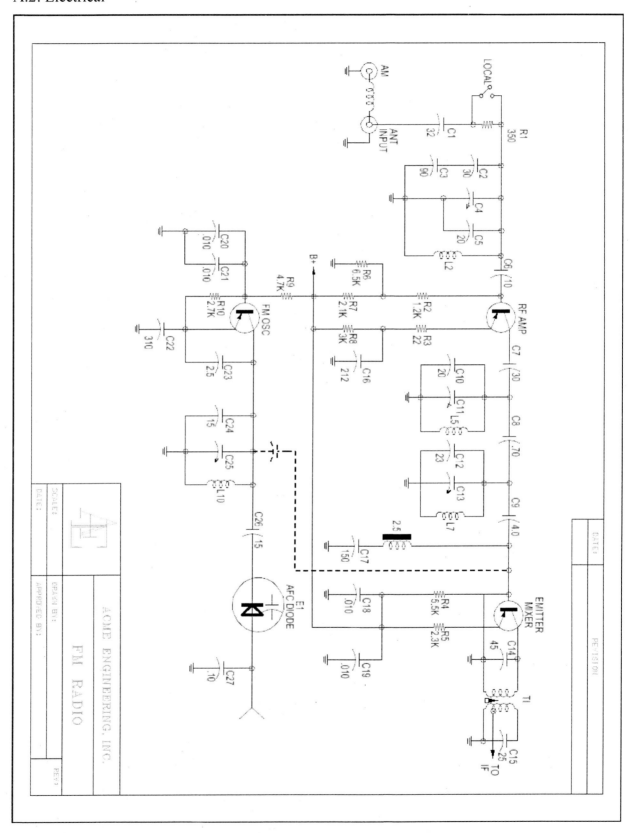

A.3: Architectural

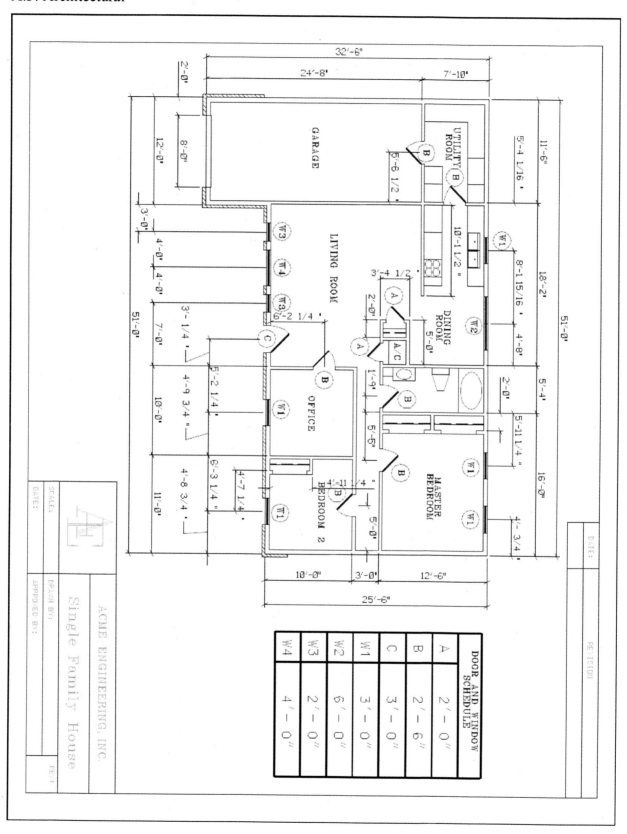

A.4: Piping

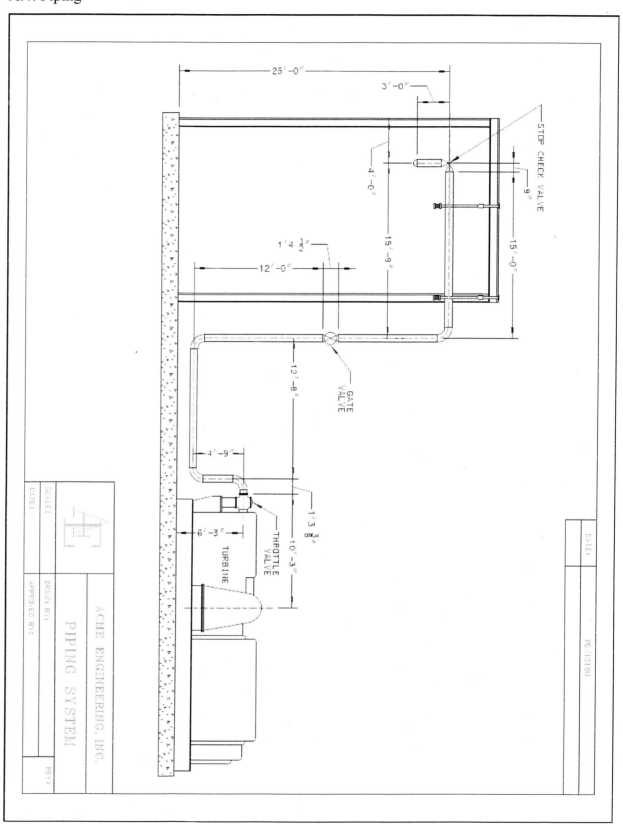

Index

To the little voice in all of us,

with love to Donna, Owen, and Allie

As I walked by myself,
And talked to myself,
Myself said unto me . . .
—Anonymous

For information address Disney Press, 1101 Flower Street, Glendale, California 91201.

Printed in Malaysia
First Edition
3 5 7 9 10 8 6 4 2
H106-9333-5-15031
Designed by Scott Piehl
Reinforced binding

Library of Congress Cataloging-in-Publication Data
Pilcher, Steve, author, illustrator.
Over there / story and art by Steve Pilcher.—First edition.
pages cm.—(Pixar Animation Studios artist showcase)
Summary: When Shredder, a little shrew who lives alone, overcomes his worry and sets out to explore what lies beyond the forest, he finds himself in trouble and discovers a new friend.
ISBN 978-1-4231-4793-0
[1. Shrews—Fiction. 2. Adventure and adventurers—Fiction. 3. Moles (Animals)—Fiction. 4. Friendship—Fiction.] I. Title.
PZ7.P62820ve 2014
[E]—dc23 2013050388

Visit www.DisneyBooks.com

PIXAR ANIMATION STUDIOS ARTIST SHOWCASE

Over There

STEVE PILCHER

DISNEP PRESS

New York · Los Angeles

It
wasn't
that
the
forest
was
too
quiet,

or
that
the
insects
were
too
noisy.

Shredder's bed was cozy enough under
the roots of an old maple...

...and
there
was
plenty
to
eat

for a little shrew.

It's just that,

well...

for a start,
his pet acorn
couldn't
talk.

And when he swung on a rubber band
and sang a little song, there was no one
to sing along.

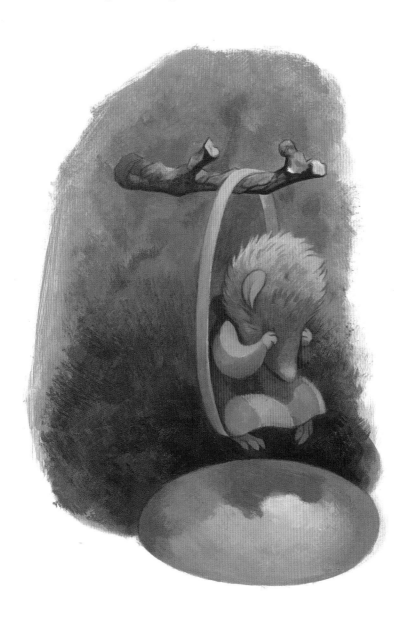

"Is there...could there...be something more?"

he asked himself.

From his lookout, Shredder could see
a shiny object sparkling in the distance.
Perhaps...he thought, there's an answer...
over there.

But what if something bad happens?

He worried most of the night. But
the more Shredder thought of the
shiny object, the more curious he got.

So in the morning, he cleaned up,
dressed, and left his home to explore.

On
the
way,
the
sky
got
bigger,
and
the
grass
got
taller.

Climbing was a little bit scary.

Crawling was much safer.

As he shuffled through the grass, Shredder heard a trickling, bubbly sound. It was a stream of water. He wanted to go to the other side, but he could not cross.

Suddenly,
the
shiny
object
caught
his
eye.

It
was
a
tiny
silver
boat.

It
floated
perfectly
for a
minute...

... then
quickly
filled
with
water
and
sank.

And just when all seemed lost,

someone or something grabbed him by the tail!

"My name is Nosey. What's yours?"
asked a dark, furry mole.

"Shredder," the shrew whispered.
"I really am a very good swimmer, you know.
I'm just a little out of practice."

"I'm a professional digger," said the mole.
"Wanna dig?"

"Sure," said Shredder.

And off they went.

When
a
giant
shadow
came
over
them,
they
hid
inside
a cave.

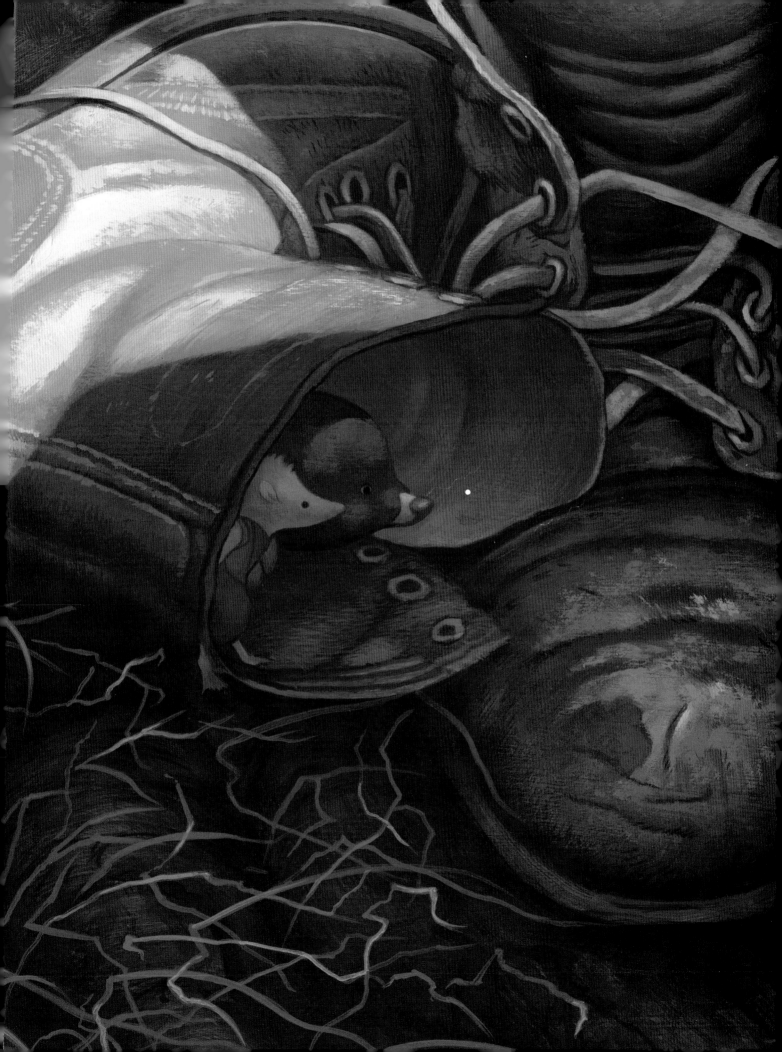

"That was scary," said Nosey.

Shredder nodded and looked all around.
The sunlight hurt his tiny eyes.
He missed the forest.

"Maybe together we can find your home.
Do you have any tasty snacks?" Nosey asked.

"I think so," replied Shredder. Then he tugged
at his clothes and pointed to Nosey's suit.
"Did you know that red and blue make purple?"

"That's my favorite color," said Nosey.

And as they walked and talked,
somehow...
everything
seemed
better.

It
wasn't
that
the
forest
was
so
nice
and
quiet,

or
that
the
insects
filled
the
air
with
music.

It's just that, with a friend,
you can play pretend.
And if you sing a little song,
your friend can sing along.

The End